Galact1c

THE ULT1MATE EXPER1ENCE

PREPARE TO LAUNCH....

INTO THE ULTIMATE JOURNEY!

CONTENTS

FOREWORD BY SIR RICHARD BRANSON

VIRGIN HAS BEEN THE STORY OF MY LIFE. Right from the start, I tried to make the brand a force for positive change, with the simple aim of creating better and happier lives. Things that are really hard but also really worthwhile, and actually mean something, tend to bring out the absolute best in people. I think that comes down to pride. I have always believed, and have seen a thousand times in business, that if your people are proud of their company's product, then everything else falls into place. And in a typically perverse, human way, the greater the challenge, the greater the determination and the greater the pride. That, in essence, is why I believe that Virgin Galactic is not only the ultimate Virgin business, but will also be the greatest and most successful business we have ever launched. As I write, the world's first private commercial passenger spaceships (yes, spaceships!) are built and flying in a fantastically successful test program; we have a truly awe-inspiring spaceport (another first) awaiting us, and, most excitingly, hundreds of aspiring Future Astronauts from around the world, ready to experience the life- changing force of space for themselves. I am inspired on a daily basis by our space project, not least because I see how it inspires others. Yes, of course, I cannot wait to get there myself, but in 40 years of business, I have never seen anything create so much excitement and inspiration, so much hope, optimism, and belief in the future. This is despite, or maybe because, what we are doing at Galactic has never been done before—and it is hard. I spend a lot of my own time now, working on and supporting Virgin initiatives that go way beyond the old and unsustainable model of growth and profit regardless of true cost. As I do that, I see that the challenges we face in the next decades are immense, and sadly, there are no silver bullets. However, the ultimate experience that Virgin Galactic represents, both literally and figuratively, gives cause for great optimism. To a generation that has started to believe that the latest edition of a smartphone represents the peak of human achievement, we are re-firing the true spirit of exploration, adventure, and discovery. And by transforming the cost, the safety, the environmental impact, and just about everything else associated with space travel of the past, we can truly help build a future fit for the next generations. Virgin Galactic started with a wonderful dream of an ultimate experience. Daring to dream can lead to outcomes beyond imagination. Welcome to our incredible journey.

"IF YOU DON'T DREAM, NOTHING HAPPENS."

Sir Richard Branson, Founder, Virgin Galactic

BOOKING A PLACE **IN HISTORY**

DREAM OF FLIGHT

THE DESIRE TO TAKE TO THE SKIES
has preoccupied humans throughout
history and across almost every culture.
From ancient myths to pioneering
balloonists and experimental gliders,
flight remained a persistent but elusive
dream rather than a practical reality.
It wasn't until the arrival of the gas engine
in the 20th century that powered flight
finally took off.

c.468ʙᴄᴇ	c.400ʙᴄᴇ	1250ᴄᴇ	1485	1709	1783	1785	1797
Mo Di, a famous Chinese philosopher, invents the kite. It was an eagle made of wood	The legend of Daedalus and Icarus, who flew too close to the Sun, is recorded by Apollodorus	Roger Bacon, English cleric and experimental scientist, first writes about mechanical flight	Leonardo da Vinci, Renaissance artist and inventor, designs a flying machine	Brazilian priest, Bartolomeu de Gusmão, known as "flying man" designs a model glider	First flight of the Montgolfier brothers' hot-air balloon takes place on June 4 in France	Balloonists Jean-Pierre Blanchard and John Jeffries cross the English Channel	French balloonist, André-Jacques Garerin makes the first parachute jump from a balloon

"WHAT FREEDOM LIES IN FLYING! WHAT GODLIKE POWER IT GIVES TO MAN!"

Charles A. Lindbergh, American aviator

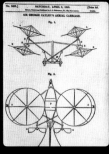

c.1843

English engineer, George Cayley designs and builds a biplane, in which a 10-year-old boy flew.

1852

First flight of French engineer Henri Giffard's steam-powered airship, from Paris to Trappes.

1895

Known as the world's first aviator, German aviation pioneer Otto Lilienthal flies a biplane glider.

1901

Alberto Santos-Dumont, Brazilian aviator, circles the Eiffel Tower in an airship in less than 30 minutes.

1903

The Wright brothers, Orville and Wilbur, make the first powered, controlled flight in the *Wright Flyer*.

1909

Louis Blériot, French aviator, makes the first crossing of the English Channel in an airplane.

1926

American inventor, Robert H. Goddard, builds and tests the world's first ever liquid-fueled rocket.

1927

Charles A. Lindbergh, US aviator, crosses the Atlantic in the *Spirit of St Louis*.

"TO GO PLACES AND DO THINGS THAT HAVE NEVER BEEN DONE BEFORE—THAT'S WHAT LIVING IS ALL ABOUT."

Michael Collins, astronaut on Gemini 10 and Apollo 11

     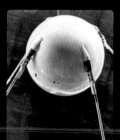  

1930

Frank Whittle, RAF engineer and British inventor, patents his design for the turbojet engine.

1932

American aviator, Amelia Earhart, is the first woman to fly solo nonstop across the Atlantic.

1939

The first fully jet-propelled aircraft, the German *Heinkel He 178*, takes its first flight.

1944

The *V-2*, the first long-range ballistic missile, developed by the Nazis, is the first modern rocket.

1947

The *Bell X-1*, piloted by Charles E. Yeager, is the first aircraft to fly faster than the speed of sound.

1957

The first man-made satellite, called *Sputnik*, is launched by the Soviet Union.

1958

The Boeing 707 jet airliner goes into commercial service, marking the dawn of the jet age.

1961

Yuri Gagarin, a Soviet cosmonaut, becomes the first man to ever go to space.

SpaceShipTwo glides back to Earth.

1969

Apollo 11 Moon landing—Neil Armstrong and Buzz Aldrin are the first humans to set foot on the Moon.

1970

The Boeing 747 enters commercial service. It more than doubles the passenger capacity of its predecessors.

1971

The first space station of any kind, *Salyut 1*, is launched into orbit by the Soviet Union.

1976

Supersonic passenger aircraft, *Concorde*, makes its first commercial flight. There are 100 passengers.

1981

The US launches its shuttle *Columbia*, which is the first partially reusable spacecraft.

1998

The *International Space Station* is launched into orbit, 220 miles (354 km) above the Earth.

2004

SpaceShipOne, piloted by Brian Binnie and Mike Melvill, wins the Ansari X PRIZE.

2013

SpaceShipTwo, based on SpaceShipOne's design, completes its first rocket powered flight.

AN ICON IS BORN

UNVEILED IN 2009, Virgin Galactic's SpaceShipTwo takes its place in a lineup of aviation icons. The spacecraft bears the "DNA of flight" graphic that illustrates (right, bottom to top) Icarus, the *Wright Flyer*, the *Spirit of St. Louis*, the *Bell X-1*, the Boeing 747, the *Apollo Lunar Module*, SpaceShipOne, and SpaceShipTwo.

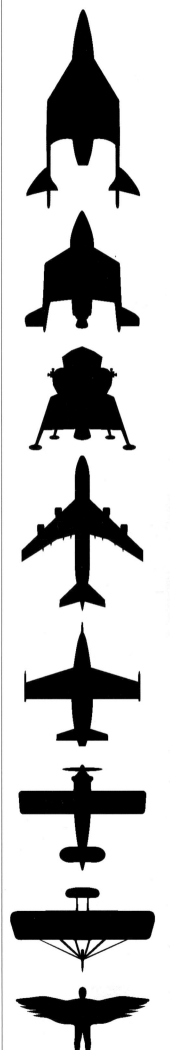

SPACESHIPTWO

SPACESHIPONE

APOLLO LUNAR MODULE

BOEING 747

BELL X-1

SPIRIT OF ST. LOUIS

WRIGHT FLYER

ICARUS

"SHE FLEW LIKE A BIRD. ONLY FASTER."

Alvin M. "Tex" Johnston, Boeing test pilot on the maiden flight of the Boeing 707

BOEING 707-320B *Intercontinental*

JET SET

Early jet travel was still considered a glamorous luxury, epitomized by the Pan Am air stewardesses of the late 1950s and early 60s.

SHRINKING THE WORLD

The Boeing 707 was the first commercially successful jetliner. It could carry up to 189 passengers across the Atlantic without stopping to refuel.

THE JET AGE

INTERCONTINENTAL TRAVEL BECAME A POSSIBILITY for millions with the introduction of jet aircraft. Powered by turbine engines, jetliners could travel faster and higher than earlier aircraft, cutting journey times dramatically. Larger models, like the Boeing 707, could carry many more passengers, making long-distance routes commercially viable with lower ticket prices. This ushered in a new opportunity for people from a wider range of social classes.

"THAT'S ONE SMALL STEP FOR [A] MAN. ONE GIANT LEAP FOR MANKIND."

Neil Armstrong, stepping on to the Moon

APOLLO 11 ASTRONAUTS NEIL ARMSTRONG MADE HISTORY ON 21 JULY 1969.

MOON LANDING

ONE MAN'S SMALL STEP was the culmination of years of planning by a team of thousands. The successful Apollo 11 Moon landing on July 21, 1969 saw the end of the Space Race between the United States and Soviet Union. The battle led to technological advances and an unparalleled desire to explore our universe. The live broadcast of the event caught the public imagination in a way that has rarely been matched.

WALKING ON THE MOON

US astronaut Neil Armstrong descended the steps of the *Eagle* lunar module to become the first person to walk on the surface of the Moon.

Mission to the Moon

Apollo 11 aimed to fulfill former US President John F. Kennedy's goal, set out in 1961, of putting men on the Moon and returning to Earth.

1 LIFT OFF
The rocket launched from Cape Kennedy on July 16, 1969, reaching trans-lunar orbit in under 3 hours.

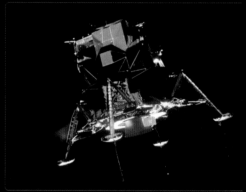

2 EAGLE LANDING
The lunar module, *Eagle*, detached from command module Columbia, and landed on the Moon.

IN NUMBERS:

MISSION DURATION:
8 days, 3 hours, 18 min, 35 sec

TIME SPENT ON THE MOON:
21 hours, 31 min, 20 sec

HOW CLOSE THE LANDING MODULE CAME TO RUNNING OUT OF FUEL:
approximately 40 seconds

LENGTH OF TIME THE RETURNING ASTRONAUTS WERE KEPT IN QUARANTINE:
3 weeks

3 LANDING AT SEA
Reentering Earth's atmosphere on July 24, the astronauts parachuted from Columbia into the Pacific Ocean.

GOING SUPERSONIC

ADVANCES IN AVIATION TECHNOLOGY made it possible for aircraft to travel faster than the speed of sound. The *Bell X-1* made the first supersonic flight in 1947, but this technology was not applied to passenger aircraft until the arrival of Concorde in 1969. The joint British–French venture promised to revolutionize long-distance flight by cutting traveling times drastically. It was not a commercial success, however, and was retired from service in 2003.

TECHNICAL FILE:

CRUISE SPEED: Mach 2.04

CRUISE ALTITUDE: 60,000ft (17,700m)

HOURS OF TEST FLIGHTS:
5,335 (2,000 were supersonic)

AVERAGE LENGTH OF TRANSATLANTIC FLIGHT: 3.5 hours

"YOU CAN **BE IN LONDON** AT **10 O'CLOCK** AND IN **NEW YORK** AT **10 O'CLOCK**. I HAVE NEVER FOUND ANOTHER WAY TO BE IN **TWO PLACES AT ONCE.**"

Sir David Frost, British journalist and Concorde regular

TIME TRAVELER
Concorde cut the flight time from London to New York in half, but the high ticket price meant that it remained an option only for the privileged elite.

TECHNOLOGICAL TRIUMPH
Many of the aviation breakthroughs seen in Concorde's design, such as the swept-back delta wings (right) for more efficient flight and drooping nose (far right) for improved landing visibility, are used as standard in aircraft design today.

LOFTY AMBITIONS
Richard Branson attempted to buy British Airways' fleet of five Concordes to keep the planes flying, but his offers were refused.

EXPLORING SPACE

AFTER THE CLAMOR OF THE SPACE RACE, the progress of space exploration slowed. Once conquered, the Moon held little further interest for the United States so NASA's last lunar mission took place in 1972. An era of cooperation between the US and the Soviet Union saw joint missions, in which US shuttles visited the Soviet space station. Later, their collaboration led to the establishment of the International Space Station (ISS).

Space missions

While manned spaceflights continued with NASA's Space Shuttle Program and the Soviet *Soyuz* spacecraft, efforts to explore further afield saw the development of unmanned probes. Probes, which cost less and are less risky than manned spaceflights, are sent to more distant planets that would take too long for crewed flights to reach.

HUMANITY'S HOME IN ORBIT

An astronaut carries out vital maintenance on the International Space Station (ISS). The product of five space agencies (those of the USA, Europe, Russia, Japan, and Canada), the ISS has been continuously inhabited for more than 13 years, by astronauts from 15 different countries.

1961: *VOSTOK*
Spacecraft piloted by the first man to go into space, Yuri Gagarin, for the Soviet Union

1981: *STS-1 COLUMBIA*
The first reusable spacecraft was NASA's space shuttle. The programme ran for 30 years.

1997: *MARS PATHFINDER*
Pathfinder took the robotic *Sojourner* to Mars to sample and photograph the surface.

1998: *INTERNATIONAL SPACE STATION (ISS)*
Habitable, artificial satellite in lower Earth orbit

2004: *CASSINI-HUYGENS*
Unmanned spacecraft enters orbit around Saturn. It launched in 1997 and its mission is ongoing.

"SPACE TRAVEL IS THE ONLY TECHNOLOGY THAT IS MORE DANGEROUS AND MORE EXPENSIVE NOW THAN IT WAS IN ITS FIRST YEAR."

Burt Rutan, Founder, Scaled Composites

BUILDING **THE DREAM**

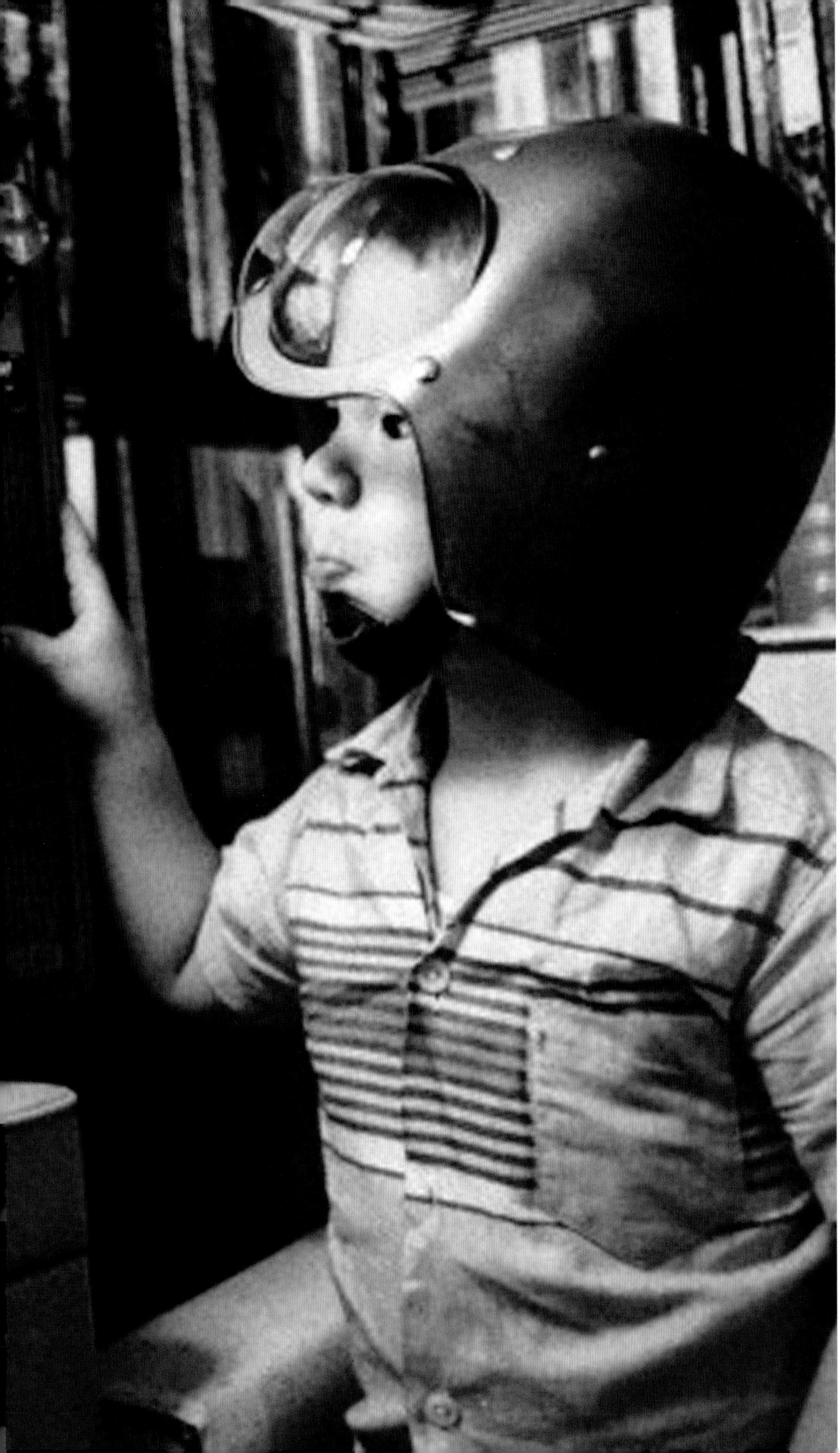

> ## "IT'S BEEN A CHILDHOOD DREAM OF MINE, EVER SINCE I SAW THE MOON LANDING, TO GO TO SPACE."
>
> Sir Richard Branson, Founder, Virgin Galactic

A CHILDHOOD DREAM

• *July 21, 1969, 2:26am*

All around the world, people watched astronaut Neil Armstrong walk on the Moon. The Apollo 11 mission inspired many young people, like the boy pictured here, to dream that one day they too would visit space. Richard Branson was no exception. Watching the event on television with his family in Scotland, the seeds of Virgin Galactic were sown. He is now making the dream of space travel come true not only for himself, but also for hundreds of others.

PROFILE: SIR RICHARD BRANSON

AFTER BECOMING AN ENTREPRENEUR AT THE AGE OF 16, Richard Branson was a millionaire by the time he turned 25. Born in London in 1950, he started building the Virgin brand with a record label in 1972. The Virgin Group now includes media, transport, and entertainment companies. The commercial spaceline Virgin Galactic will fulfil Richard's goal of going into space.

RICHARD BRANSON POSES IN A SPACESUIT TO PROMOTE HIS LONG-HELD DREAM OF COMMERCIAL SPACEFLIGHT.

FAMILY BUSINESS
Richard laughs with his son Sam and daughter Holly at an event at Spaceport America, in New Mexico. He plans to fly with his family on Virgin Galactic's inaugural flight.

"I SEE **LIFE** ALMOST LIKE **ONE** LONG UNIVERSTIY EDUCATION THAT I NEVER HAD. EVERY DAY I'M **LEARNING SOMETHING NEW.**"

Sir Richard Branson, Founder, Virgin Galactic

Q&A

What inspires you to start a new business?

I have always tried to be led by some degree of personal interest. Setting up new businesses is hard work and can be dispiriting at times. Having a personal enthusiasm for the subject matter definitely gives you an edge, an added determination to succeed, and just makes life a little more fun and worthwhile. I have also tried to avoid setting up businesses in sectors where the current incumbents are doing a good job—because—what's the point?! We like to identify areas where the consumer is being poorly served and where we have the skills, ideas, or just the attitude to do it better. I suspect this will give us plenty of new opportunities for many years to come!

Why is commercial spaceflight important to you?

I sometimes think that Virgin Galactic is the ultimate Virgin business and a very 21st-century representation of what it means to be Virgin. Firstly, we've never been scared away by the scale of a challenge—and it doesn't get much bigger than space! Then, I have had a long love affair with flying machines of one sort or another, and have always dreamt of going to space myself. Spaceflight definitely falls into the category of an industry serving its potential customers poorly—not just in its failure to open the experience of space to anyone other than a handful of government employees, but also in the huge benefits that cheap satellite launch or trans-continental travel via space could bring to everyone. This is a really important business and we really believe we have what it takes not only to do it better, but to provide a catalyst for incredible positive change at a global level.

How do you convince people to invest in a project so full of unknowns?

We didn't even try to do that until we had built a spaceship! We did have the big advantage of having a proven prototype and an early passionate and pioneering customer base which gave aabar, our partners, reassurance that this was a ground-floor opportunity to invest in a game-changing technology.

We have always approached the project milestone by milestone and never made any secret of the fact that we will need to be commercially successful if we are to realize the full potential of the new commercial space age.

Where do you see Virgin Galactic being in 20 years time?

I would hope that Virgin Galactic will be transporting thousands of people around the world via space as the leader in 21st-century, clean, fast, commercial long-haul transportation. I also hope we will be giving people incredibly romantic holidays in space hotels with optional excursions to low level moon orbits. Well—you have to dream!

What advice would you give to aspiring entrepreneurs?

Life's short—give it a go! Don't be afraid of failure, but do learn by it. Have fun, follow your passions and the money will follow. I also think there is a fantastic new opportunity to change fundamentally the way we do business and to take a lead in that could be a very smart decision. I believe the next great frontier (other than space!) will be where the boundaries between work and purpose merge into one. Where doing good, really is good for business.

Did your family take much convincing to sign up for the first flight?

I still haven't convinced my wife! She wisely says that at least one member of the family needs to keep their feet on the ground. My children share my love of adventure and took little persuading. However, each of us recognizes the scale of what we're setting out to do and I think that demonstrates the confidence we have in the space system's design and to the whole approach we've taken to development and testing, with safety always being the number one priority.

BALLOONING

DRIVEN BY A WILL TO SUCCEED, an adventurous spirit, and a sense of fun, Richard Branson has made several world record attempts. Starting with boats, he soon moved on to balloons. In 1987, he made the first crossing of the Atlantic in a hot-air balloon with balloonist Per Lindstrand. They went on to cross the Pacific, and then tried to circumnavigate the globe (see opposite). These attempts brought Richard into contact with Steve Fossett, the adventurer with whom he would collaborate on the GlobalFlyer (see pp.36-37).

" ADVENTURES ARE PHYSICALLY CHALLENGING, AND MENTALLY CHALLENGING, AND TECHNOLOGICALLY CHALLENGING, AND THAT IS WHAT MAKES THEM FASCINATING. "

Sir Richard Branson, Founder, Virgin Galactic

Around the world

Sir Richard Branson has made three attempts to circumnavigate the globe in a balloon without stopping. The experience informed his future projects, and introduced him to some of the people with whom he would work on the GlobalFlyer.

FIRST ATTEMPT
Sir Richard Branson in Marrakech, Morocco, preparing for his first attempt at circumnavigation, in 1996.

TRYING AGAIN
Richard's second effort ended when a gust of wind lifted his balloon and it took off without him in Marrakech in December 1997.

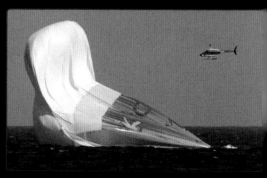

FINAL EFFORT
Richard's third, and final, attempt ended in the Pacific near Honolulu, Hawaii, on Christmas Day 1998 after drifting off course. He was rescued from the capsule along with adventurer Steve Fossett and Swedish aeronautical engineer Per Lindstrand.

THE X PRIZE

OFFERING THE LARGEST PRIZE IN HISTORY,
the $10 million Ansari X PRIZE was launched in 1996.
Modeled on aviation contests of the early 20th century,
it challenged teams to build a reusable spacecraft.
Chairman of the X PRIZE Foundation, Dr. Peter Diamandis,
set up the prize to spur innovation in spaceflight.

IN NUMBERS:

AIM OF ANSARI X PRIZE COMPETITION:
take 3 people, 62 miles (100 km) high into
space, twice within 14 days

**NO. OF TEAMS ENTERED INTO ANSARI
X PRIZE:** 26 (from 7 countries)

ANSARI X PRIZE FUND: $10 million

NO. OF X PRIZES TO DATE:
5 awarded (a further 4 are currently open)

**TOTAL PRIZE FUNDS AWARDED
TO WINNERS TO DATE:** $54 million

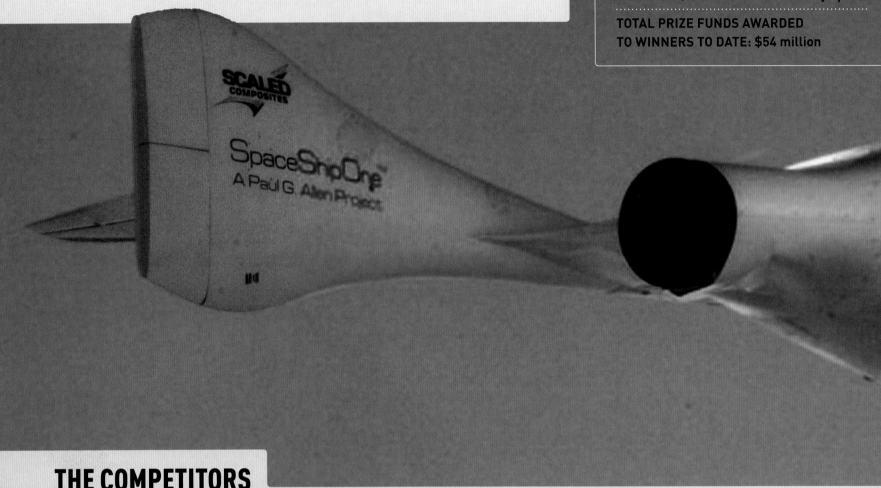

THE COMPETITORS

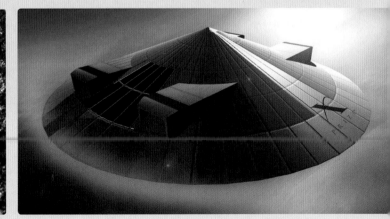

ASCENDER

Bristol Spaceplanes, a team from the UK, designed
a suborbital airplane that takes off conventionally,
but deploys rockets for the steep climb to space.

NEGEV 5

IL Aerospace Technologies
from Israel used a hot-air
balloon for takeoff.

THE SPACE TOURIST

US team, Discraft Corporation, designed a flying saucer capable of
horizontal takeoff from a conventional runway. The craft was 98ft (30m)
in diameter and ascended by sucking in and expelling air for propulsion.

SPACESHIPONE

Designed by Burt Rutan, Scaled Composites' spaceship was lifted to altitude by a carrier craft, White Knight, for an air launch, meaning it needed to carry less fuel.

MICHELLE-B

TGV Rockets' Modular Incremental Compact High Energy Low-Cost Launch Example was a vertical launch reusable rocket.

COSMOS MARINER

Texan team, Lone Star Space Access Corporation's craft took off from a coastal airport and ascended over the sea.

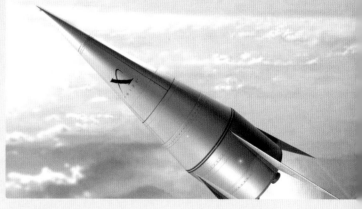

LUCKY SEVEN

The entry from US team Acceleration Engineering was a land-launched rocket designed to deploy a parafoil on descent, guided to land by global satellite positioning (GPS).

"I THOUGHT IT WOULD BE EITHER **DISASTROUS** OR A DODDLE... THERE'S BEEN **A LOT OF DRAMA.**"

Sir Richard Branson, Founder, Virgin Galactic, on GlobalFlyer's world-record breaking flight

FLYING HIGH

GlobalFlyer left Salinas, Kansas on February 28, 2005, and landed victorious on March 3. Its carbon-composite frame and twin fuselage would inform Rutan's design for White Knight.

FLIGHT PROFILE:

DURATION OF FLIGHT:
2 days, 19 hours, 1 min

FLIGHT SPEED:
342.2mph (590.7kmph)

DISTANCE FLOWN:
22,936 miles (36,912km)

CRUISING ALTITUDE:
45,000–52,000ft (14–16km)

GLOBALFLYER

IN A BID TO MAKE THE FIRST SOLO nonstop around-the-world flight, Richard Branson once again teamed up with adventurer Steve Fossett (see p.32). Richard commissioned Burt Rutan's aerospace company Scaled Composites to build the Virgin Atlantic GlobalFlyer. While working with Burt in 2002, Richard discovered that he was building a spaceship in a bid to win the Ansari X PRIZE.

PROFILE:
FOUNDER, SCALED COMPOSITES
BURT RUTAN

AEROSPACE LEGEND Burt Rutan was born in Oregon in 1943. After graduating in aeronautical engineering, he worked for the US Air Force as a flight test engineer. In 1974, he formed the Rutan Aircraft Factory, which developed *Voyager*, the first aircraft to circle the globe nonstop. He set up Scaled Composites in 1982 to develop research craft. The company has built some of the most innovative aircraft ever flown.

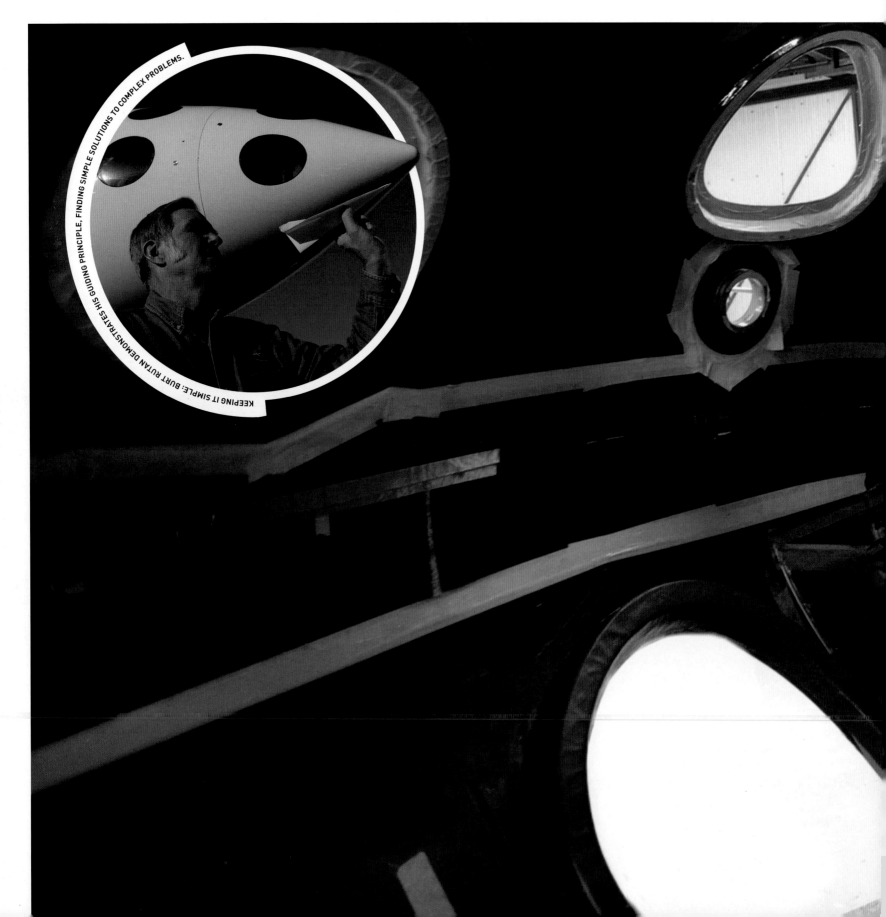

KEEPING IT SIMPLE: BURT RUTAN DEMONSTRATES HIS GUIDING PRINCIPLE, FINDING SIMPLE SOLUTIONS TO COMPLEX PROBLEMS.

"TESTING LEADS TO FAILURE, AND FAILURE LEADS TO UNDERSTANDING."

Burt Rutan, Founder, Scaled Composites

Q&A

Why is commercial spaceflight important to you?
Clearly there is an enormous pent-up hunger to fly in space, and not just dream about it. We do want our children to go to the planets. We are willing to seek breakthroughs by taking risks. And if the business-as-usual space developers continue their decades-long pace they will be gazing from the slow lane as we speed into the new space age.

What was your chief motivation in taking on the Tier One Project to win the Ansari X PRIZE?
The project was the fulfilment of a dream. Back in 1994 I started toying with what it would take for us to build a supersonic, even hypersonic airplane, the dream of this was such a phenomenally exciting ride—to go 3.5 Gs straight up! That's a neat ride! And because of Tier One, it's no longer insane to think of a private company doing a manned spaceflight.

Where do you think space travel will take us next?
Sub-orbital spaceflight was Tier One. Next will be a spaceship that goes to Earth orbit—that will be Tier Two. Anything that reaches above that, to the altitude of the Moon, or Mars, or the stars, is Tier Three.

What would you say to encourage someone to travel into space themselves?
It will be the world's greatest rollercoaster ride!

COMMERCIAL SPACEFLIGHT

THE IDEA OF SPACE TOURISM fascinated Sir Richard Branson. He visited many companies working on the technology that could make commercial spaceflight viable. Richard Branson found his answer when he discovered Burt Rutan's SpaceShipOne (see pp.48–49). By taking space travel out of government hands, Virgin Galactic aims to spur innovation in the sector.

Science fiction?
Far from a distant dream—as imagined in this picture of a futuristic spaceport—Virgin Galactic aims to send thousands of people into space in the coming decades.

"WE NEED AFFORDABLE SPACE TRAVEL TO INSPIRE OUR YOUTH, TO LET THEM KNOW THAT **THEY CAN EXPERIENCE THEIR DREAMS.**"

Burt Rutan, **Founder, Scaled Composites**

IN NUMBERS:

AVERAGE COST OF TICKET LONDON–NEW YORK IN 1958:
£173 one way (£2,914 in today's money)

COST OF TICKET ON FIRST SPACEFLIGHT:
$250,000

JOURNEY TIME LONDON–NEW YORK IN 1958:
10hr 22min

APPROXIMATE SS2 FLIGHT TIME: 2hr 30min

CRUISING ALTITUDE OF TRANSATLANTIC JET IN 1958: 40,000ft (12,000m)

"WE HOPE TO **CREATE THOUSANDS OF ASTRONAUTS** OVER THE NEXT FEW YEARS AND BRING ALIVE THEIR DREAM OF SEEING THE **MAJESTIC BEAUTY OF OUR PLANET** FROM ABOVE, THE **STARS IN ALL THEIR GLORY** AND THE **AMAZING SENSATION OF WEIGHTLESSNESS.**"
Sir Richard Branson, Founder, Virgin Galactic

ONE MAN'S DREAM

Burt Rutan's bid to win the Ansari X PRIZE, known as the Tier One project, was financed by Paul Allen of Microsoft. However, when Sir Richard Branson learned of this in 2002, he agreed to license key SpaceShipOne technology with a view to building and operating a commercial version should it be successful. Branson aims to democratize space. By developing a commercial spaceflight industry, taking it out of government hands, and introducing an element of competition, he hopes that space travel will become available to all.

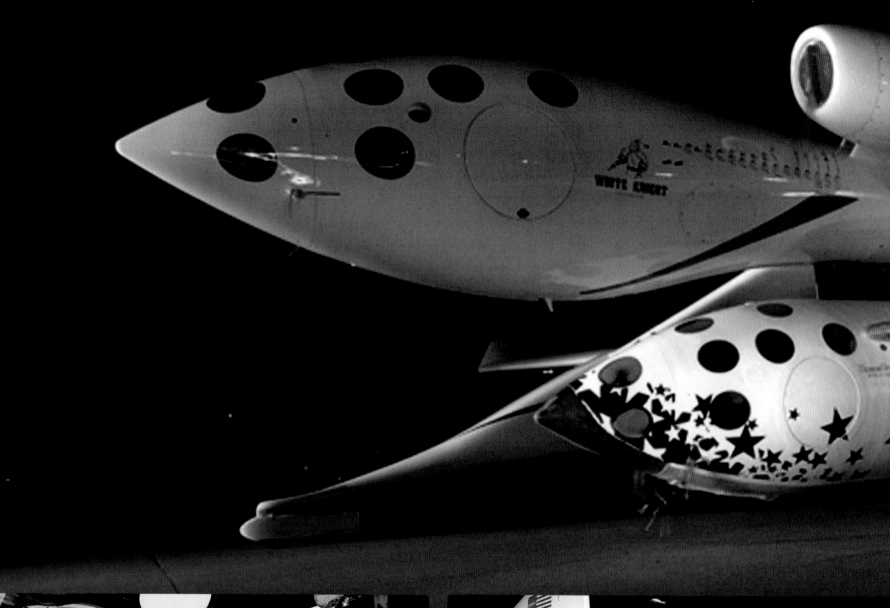

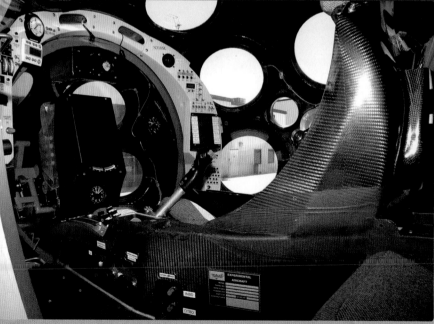

COCKPIT

The cockpit and the majority of the controls of SpaceShipOne and White Knight are identical. SpaceShipOne's cabin is 4.9ft (1.5m) in diameter and can seat one pilot and two astronauts.

ROCKET MOTOR

SpaceShipOne (SS1) awaits a fresh rocket motor. Filaments of carbon composite are woven together to form the casing for the hybrid rocket, which burns liquid nitrous oxide and rubber.

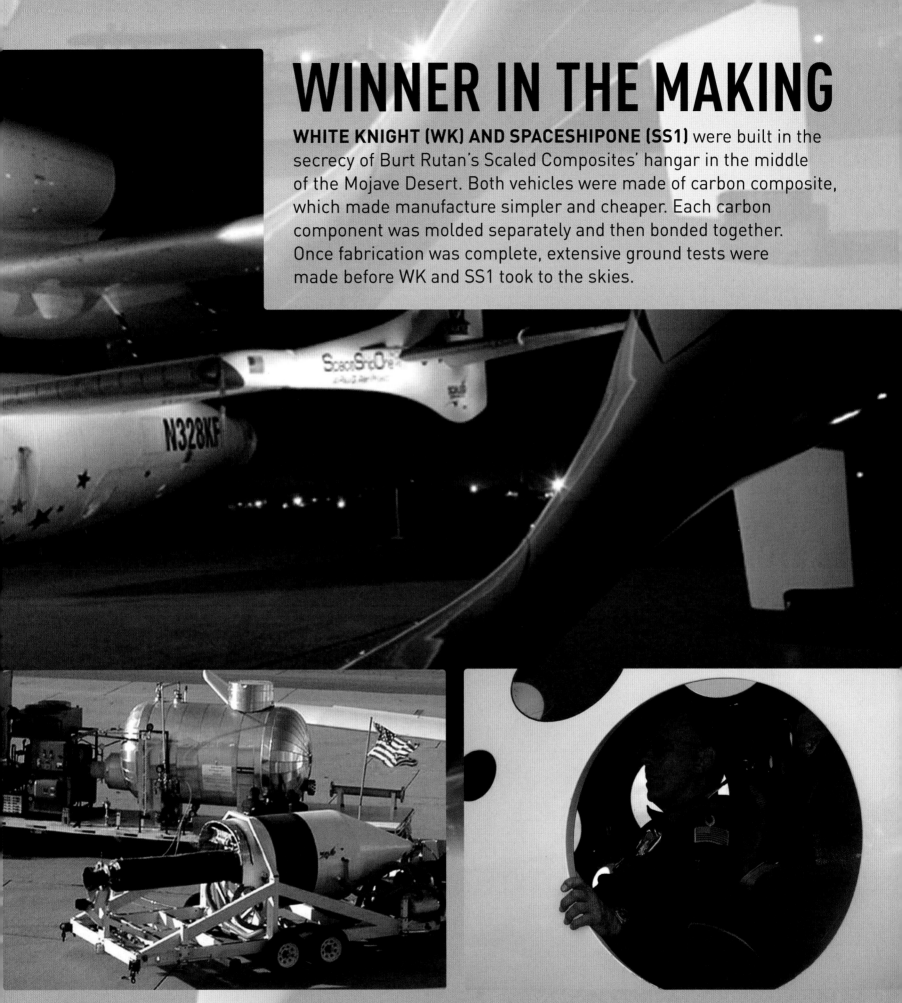

WINNER IN THE MAKING

WHITE KNIGHT (WK) AND SPACESHIPONE (SS1) were built in the secrecy of Burt Rutan's Scaled Composites' hangar in the middle of the Mojave Desert. Both vehicles were made of carbon composite, which made manufacture simpler and cheaper. Each carbon component was molded separately and then bonded together. Once fabrication was complete, extensive ground tests were made before WK and SS1 took to the skies.

TEST STAND

The hybrid rocket underwent a program of test firings on the ground to ensure its safety and effectiveness. Here the mobile nitrous oxide delivery system is being mounted on a special test stand.

READY TO ROLL

Test pilot Mike Melvill sits in the cockpit of SpaceShipOne. He flew the majority of the ship's fifteen test flights before undertaking the first of the two competitive X PRIZE flights.

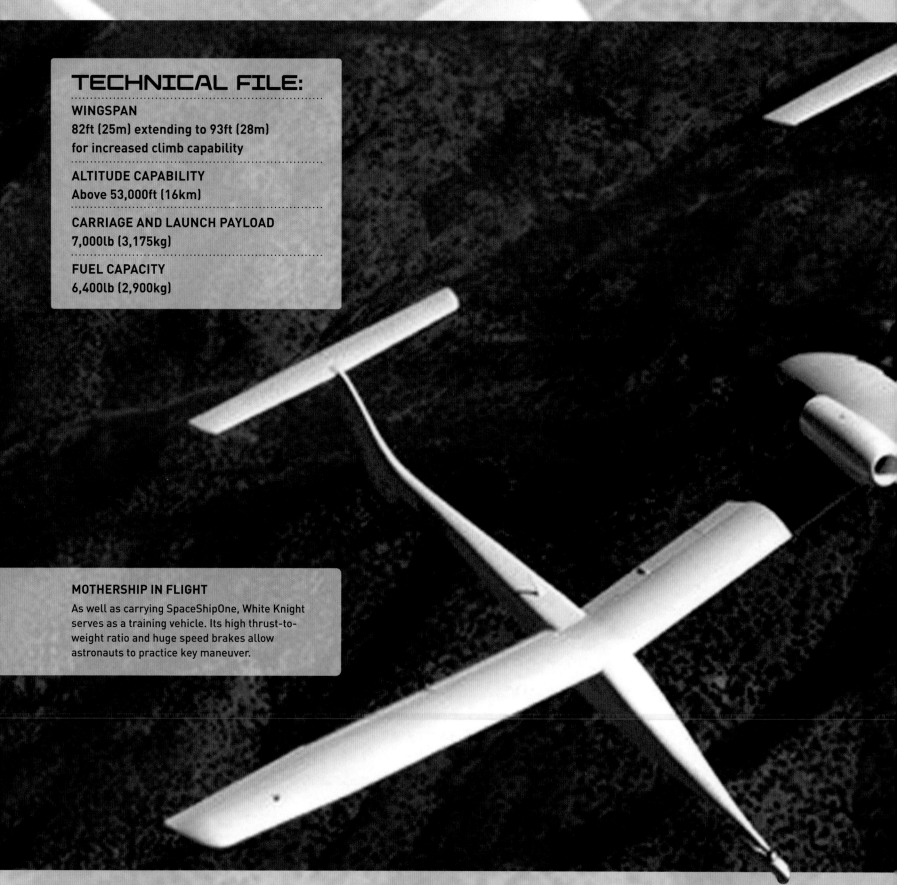

WHITE KNIGHT
THE LAUNCH AIRCRAFT

WITH AN EXCEPTIONALLY LONG wingspan, White Knight was designed to carry heavy loads to high altitudes. By providing an airborne launch for SpaceShipOne (see pp.48-49), it minimizes fuel use and increases safety. White Knight's cockpit is identical to that of SpaceShipOne, making it a realistic training environment for astronauts.

TECHNICAL FILE:

WINGSPAN
82ft (25m) extending to 93ft (28m) for increased climb capability

ALTITUDE CAPABILITY
Above 53,000ft (16km)

CARRIAGE AND LAUNCH PAYLOAD
7,000lb (3,175kg)

FUEL CAPACITY
6,400lb (2,900kg)

MOTHERSHIP IN FLIGHT
As well as carrying SpaceShipOne, White Knight serves as a training vehicle. Its high thrust-to-weight ratio and huge speed brakes allow astronauts to practice key maneuver.

SIR RICHARD BRANSON VISITS BURT RUTAN TO SEE WHITEKNIGHTTWO.

In the hangar
While collaborating on the GlobalFlyer project (see pp.36-37), Sir Richard Branson learned that White Knight was being built in secret in Scaled Composites' Hangar 78. The vast hangar accommodated the craft's 82ft (25m) wingspan, which is constructed from carbon composite.

ATTACHED
White Knight carries SpaceShipOne aloft.

"EVERYTHING LOOKS NONSENSICAL BEFORE IT WORKS."

Burt Rutan, Founder, Scaled Composites, on the design for SpaceShipOne

Tail cono
SpaceShipOne is powered by a hybrid rocket motor.
It burns non-toxic, liquid nitrous-oxide and solid rubber
fuel. The products of the burn are mostly water vapor,
carbon dioxide, hydrogen, nitrogen, and carbon monoxide.
It is less polluting than other rocket propulsion.

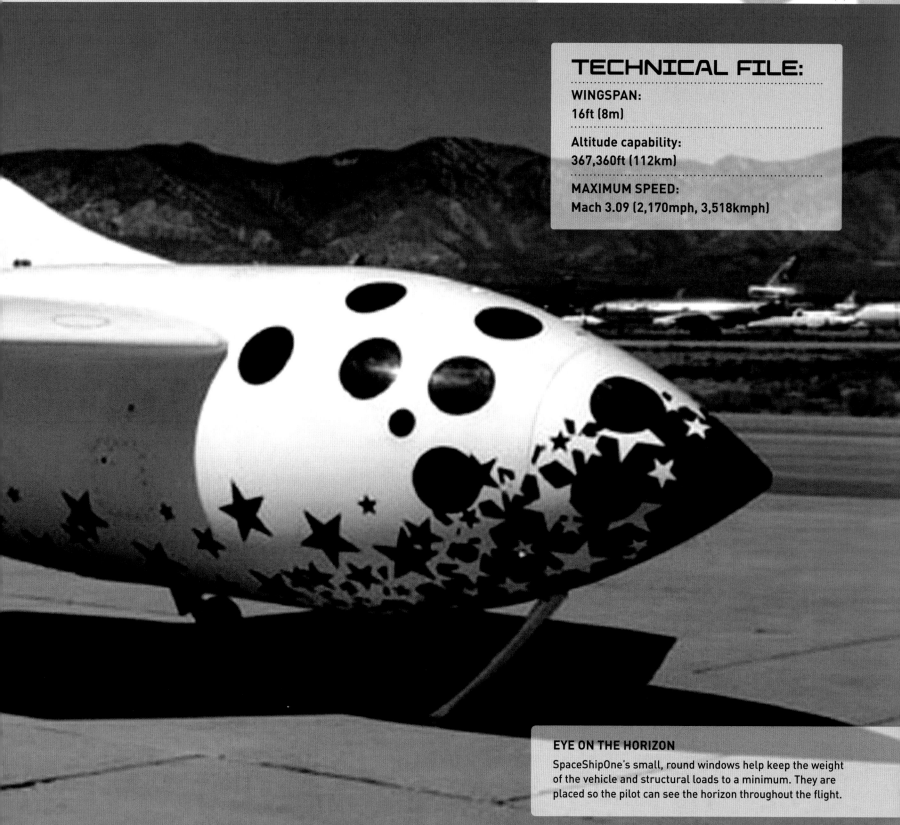

TECHNICAL FILE:

WINGSPAN:
16ft (8m)

Altitude capability:
367,360ft (112km)

MAXIMUM SPEED:
Mach 3.09 (2,170mph, 3,518kmph)

EYE ON THE HORIZON
SpaceShipOne's small, round windows help keep the weight of the vehicle and structural loads to a minimum. They are placed so the pilot can see the horizon throughout the flight.

SPACESHIPONE HEADED FOR THE STARS

DESIGNED TO CARRY THREE PEOPLE into space and meet the criteria for winning the X PRIZE, SpaceShipOne is an air-launched, carbon-composite rocket ship. Once launched, rockets fire it into space where its wings "feather", or fold up, to create high drag for reentry to Earth's atmosphere. The ship then reconfigures to glide back to the runway.

SPACESHIPONE & WHITE KNIGHT

SMALL ENOUGH TO BE MOUNTED directly under the fuselage of its carrier aircraft, SpaceShipOne (SS1) is just 28ft (8.5m) long. The mothership, White Knight, is designed to lift the 8,000lb (3,600kg) gross payload of SS1 to high altitude before releasing it. Once air-launched, SS1's hybrid rocket is fired to propel it into space.

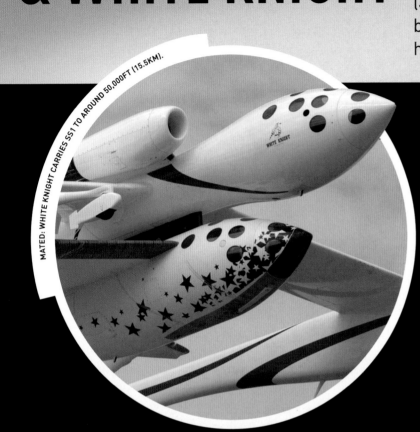

MATED: WHITE KNIGHT CARRIES SS1 TO AROUND 50,000FT (15.5KM).

TECHNICAL FILE:

CABIN DIAMETER:
White Knight and SS1: 4.9ft (150cm)

PROPULSION:
SS1: 1 hybrid rocket motor
White Knight: two J-85-GE-5 engines

SS1 ROCKET BURN TIME:
80 seconds

NO. OF COMBINED TEST FLIGHTS:
17

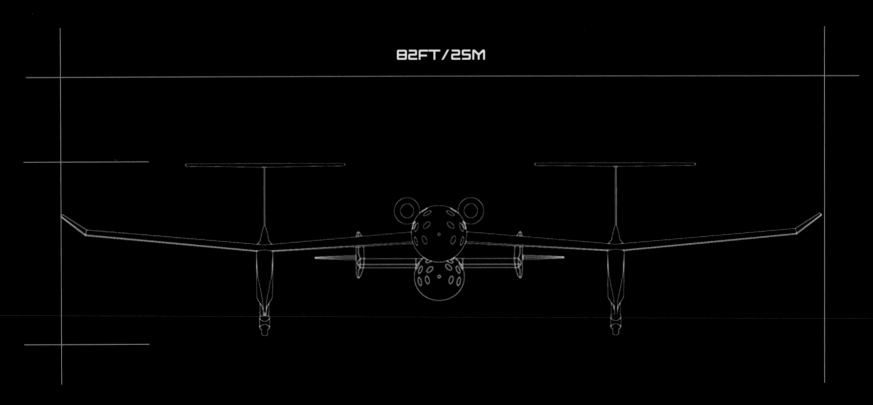

82FT/25M

"THIS IS NOT THE END BUT IT'S A VERY GOOD BEGINNING."

Burt Rutan, Founder, Scaled Composites, on SpaceShipOne

Dimensions
White Knight's w-shaped wing and small cabin leaves space for SpaceShipOne to be secured under its fuselage. The cabins of both vehicles are identical.

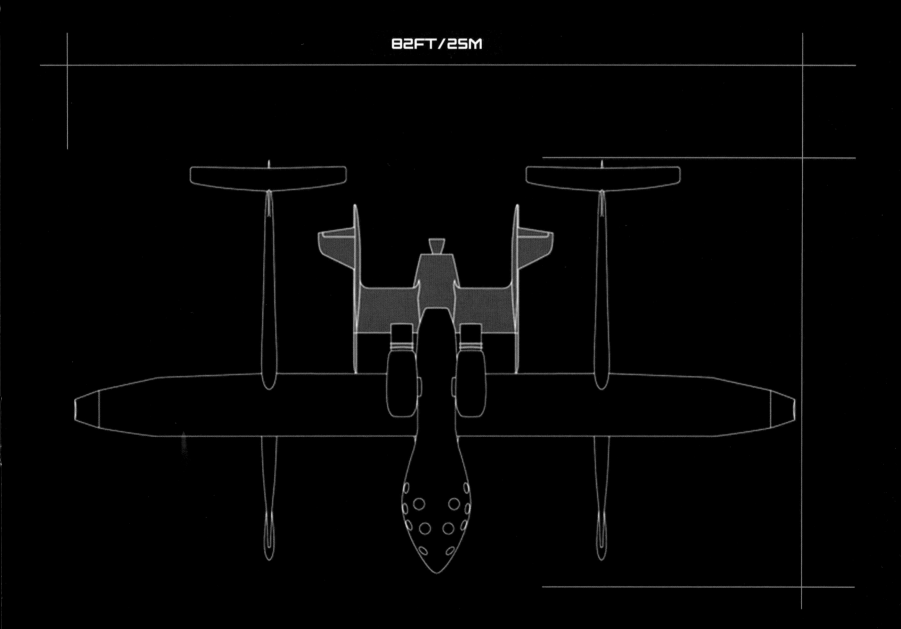

82FT/25M

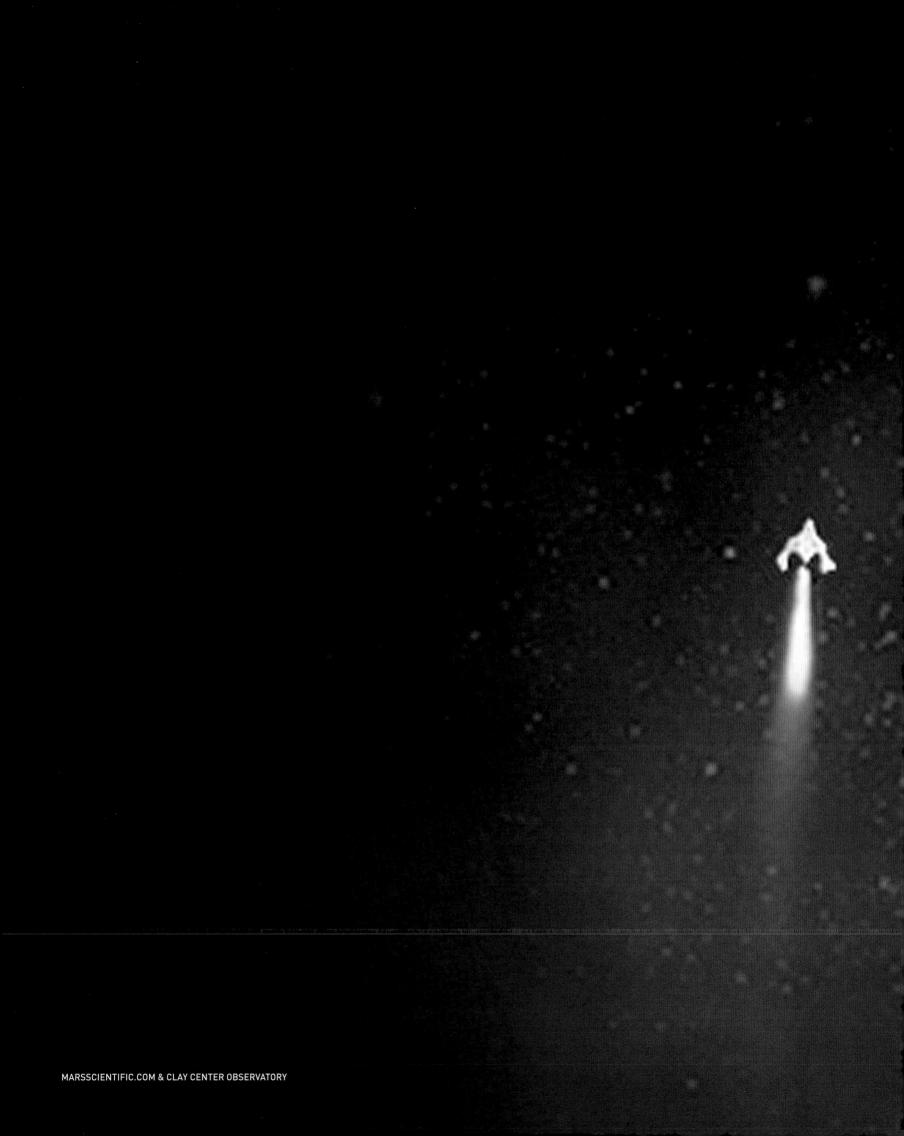

"THE SHUDDERING AND SHAKING... THE DEMONIC SCREECHING OF THAT MOTOR WERE MOST MEMORABLE."

Brian Binnie, pilot of SpaceShipOne

ROCKET BOOST
Clay Center Observatory, Massachusetts

• *October 4, 2004*

SpaceShipOne's rocket blast was captured by a powerful video-tracking telescope at the Clay Center Observatory. This image combines three frames, showing the initial blast and vertical acceleration during Brian Binnie's X PRIZE-winning flight.

PROFILE: TEST PILOT BRIAN BINNIE

AFTER GROWING UP IN SCOTLAND, Brian Binnie studied at Brown and Princeton Universities in the USA. He joined the US Navy and qualified as a naval test pilot in 1988. In 2003, working with Scaled Composites, he flew the first powered flight of SpaceShipOne on the 100th anniversary of the Wright brothers' first powered flight. His X PRIZE-winning flight earned him his Astronaut Wings.

"MY REALIZATION THAT DAY... WAS THAT IF YOU HAVE NO FEAR, THEN YOU HAVE NO DREAMS."

Brian Binnie, pilot of SpaceShipOne, on the day of the X-PRIZE-winning flight

Q&A

What goes through your mind in the moments before beginning a test flight?
Being a Navy guy, I was influenced by Admiral Alan Shepard who is credited with the test pilot's prayer: "Dear God, please don't let me screw up".

Can you describe what it's like to experience g-force?
Exhilarating. The one relentlessly provided by the rocket is front to back through the chest.

What does it feel like to be in zero gravity?
Wonderful! I submit you cannot be detached, indifferent or depressed in this phase of the flight. It would be a great tonic for all that suffer from those types of afflictions.

How does it feel to see Earth from space?
Humbling and inspiring. The contrast between that "black sky" and the "blue pearl" of the atmosphere makes it difficult to imagine that our planet was just an evolutionary oddity.

Q&A

Did you always dream of flying?
Actually, it was a business requirement. I was doing a lot of traveling in airlines and came to the conclusion that I needed to get a private pilot's licence. I worked my way up to a commercial license. I thoroughly enjoyed all aspects of flying and wanted to do it as much as possible.

What is it like to experience G-force?
With the rocket thrust on the way up, it's very disorienting as you have downward g of about 3.5 and about 3gs on your back, and that combination makes you feel like you're doing a loop. You get this sensation in your mind that you're going to go over on your back.

Can you describe what happens when SS1 launches from White Knight?
It's quite a shock—we just drop off. The White Knight pilot pulls a handle—it's a mechanical release. You immediately light the rocket motor, as you're falling like a brick, and accelerate to the speed of sound in 9 seconds. So, you have 3gs on your back and it gets more as it goes because it's burning fuel. Then you have to pull back on the stick to turn the nose up and it's a pretty abrupt turn because you don't want to be wasting thrust going horizontal. That first 10-12 seconds is real critical for the pilot—it's the hardest thing to do as there's a lot going on.

PROFILE: TEST PILOT MIKE MELVILL

BORN IN SOUTH AFRICA, Mike Melvill moved to the USA in the 1970s. The first time they met, Mike showed Burt Rutan the VariViggen aircraft he had built at home. Melvill has flown the maiden flights of 10 of Rutan's aircraft. Mike holds nine World Air Sports Federation (FAI) records and was awarded his Astronaut Wings (featured in the above photograph) two hours after landing the first spaceflight of SpaceShipOne on 21 June 2004.

A MOMENT TO REMEMBER

DRIFTING ABOVE THE EARTH at 367,442ft (111.9km), Brian Binnie was making history in SpaceShipOne on October 4, 2004. He piloted the second X PRIZE flight and flew into the record books by reaching the highest altitude for a rocket plane. Mike Melvill had completed the first successful flight on September 29. If Binnie landed safely, the prize was theirs.

PAUSE FOR A PICTURE
SpaceShipOne • *October 4, 2004*

Far above the world, Brian Binnie takes a moment to look back at Earth from the cockpit window and uses his camera to record the view for posterity before making his descent.

"IT WAS **A THRILL** THAT I THINK EVERYBODY SHOULD HAVE **ONCE** IN A LIFETIME."

Brian Binnie, pilot of SpaceShipOne, on the final X PRIZE flight

"WHAT BEAUTY. I SAW CLOUDS AND THEIR LIGHT SHADOWS ON THE DISTANT DEAR EARTH..."

Yuri Gagarin, Russian cosmonaut, on seeing Earth from space

WONDERFUL WORLD

• *October 4, 2004*

This view of the Earth from space was taken by pilot Brian Binnie from the cockpit window of SpaceShipOne during the second competitive X PRIZE flight. The curvature of the Earth is clearly visible and the thin blue line of the atmosphere stands out against the deep black of space.

WINNING TEAM

The Tier One team celebrates winning the X PRIZE. From left to right: X PRIZE Chairman Dr. Peter Diamandis, financier Paul G. Allen, designer Burt Rutan, pilot Brian Binnie, and Virgin Galactic founder, Richard Branson.

HEIGHT OF SUCCESS: X PRIZE-WINNING PILOTS BRIAN BINNIE AND MIKE MELVILL.

STARTING WITH A WIN

AS SPACESHIPONE TOUCHED DOWN on the Mojave runway, the Tier One team won the X PRIZE. With this historical success, Virgin Galactic was born. Richard Branson agreed to license the technology developed by Paul G. Allen's Mojave Aerospace Ventures—the company backing Tier One—to build a second-generation vehicle for commercial spaceflight. Branson had registered the name Virgin Galactic back in 1999 on a whim—now his dream was becoming a reality.

"IT [SPACESHIPONE] WILL INSPIRE THE NEXT GENERATIONS..."

General John R. Dailey, Director of the Smithsonian National Air and Space Museum

STAR OF THE SHOW
Oshkosh, Wisconsin
• *July 25, 2005*

SpaceShipOne and White Knight draw a crowd at the EAA Airventure airshow. Shortly after the show, Mike Melvill took the vehicles on their final flight to the Smithsonian Institute's National Air and Space Museum, where they are on displayed alongside the *Spirit of St. Louis*, the *Bell X-1*, the *Wright Flyer*, and *Columbia* space shuttle.

THE EARLY DAYS

FROM HUMBLE BEGINNINGS at a spare desk at Virgin Group's offices, Virgin Galactic's first employee, Stephen Attenborough has helped build a spaceline. He initially employed a team of five whose first task was to set up a company website with little more than a logo, footage of the X PRIZE flight, and a form for aspiring astronauts to fill in. The response was overwhelming. They were soon receiving deposit cheques from all over the world.

IN NUMBERS:

NUMBER OF FUTURE ASTRONAUTS:
March 2010 = 330
April 2013 = 580
January 2014 = 680

TOTAL DEPOSITS BY 2014:
$80 million

NUMBER OF PEOPLE WHO HAVE REGISTERED INTEREST (AS OF FEBRUARY 2014): 85,000

"THERE WAS A DELUGE."

Stephen Attenborough,
Commercial Director, Virgin Galactic

**HIGH FLYERS
MOJAVE, CALIFORNIA**

A group of employees, including then President, Will Whitehorn, (front, second left), Stephen Attenborough (front, right), and designer Philippe Starck (back, right) pose in front of White Knight.

GALACTIC PRESIDENT
When Virgin Group's Head of Special Projects, Will Whitehorn, tried to register the Virgin brand for use in space, he found Sir Richard Branson had done so a decade before.

HALF MOON STREET
Early employees of Virgin Galactic line the balcony of their offices at the appropriately named Half Moon Street in Mayfair, London, UK.

FINDING AN IDENTITY

BEARING THE FAMILIAR BRIGHT RED, a fledgling Virgin Galactic logo first appeared on SpaceShipOne. However, the new company needed to find its own identity, and it was felt that blue was more appropriate for a spaceline. French designer Philippe Starck created a logo based on the iris of Sir Richard Branson's eye. The spacecraft's livery includes the DNA of flight icons, which place the vehicle in a line of aviation firsts.

Pushing the boundaries
The image of an iris looks back to Earth from SpaceShipTwo (SS2). Starck intended for the logo to prompt people to reflect on the basic human instinct to push boundaries and explore, stating, "the eye's pupil incorporates an eclipse, the dawning of something new."

Virgin territory
Richard Branson's mother inspired the Galactic Girl logo, which adorns the first SS2 aircraft, *VMS Eve*. The mothership is also named for her.

Seeing red
This early logo uses the Virgin Group red for a spaceship icon eclipsing the Sun, over a gray shape that resembles an astronaut's helmet.

DNA of flight
SpaceShipTwo flies solo on its own livery (right) but is attached to WhiteKnightTwo on the carrier craft (see p.77).

Visionary design
The logo on SpaceShipTwo reflects the fact that people will be able to look at Earth from space with their own eyes for the first time.

TREVOR BEATTIE
British advertising executive,
responsible for marketing
campaign for Virgin Galactic

STEPHEN ATTENBOROUGH
Commercial Director and
first full-time employee of
Virgin Galactic

FOUNDER ASTRONAUTS
Mojave, California

Following the X PRIZE victory and the birth of Virgin Galactic, there was a flurry of people eager to sign up for the first flights. Some of the earliest future astronauts traveled to Mojave to see White Knight—pictured here with Commercial Director of Virgin Galactic Stephen Attenborough and Burt Rutan (front, center).

BURT RUTAN
Founder of innovative aerospace company, Scaled Composites

PHILIPPE STARCK
Designer responsible for designing the Virgin Galactic logo

TEAM EFFORT

The Spaceship Company (TSC) employees line the hangar as a completed WK2 and SS2 are rolled out. The hangar doors comfortably accommodate WK2's 140ft (43m) wingspan.

BUILDING SPACESHIPS

THE FORMATION OF THE SPACESHIP COMPANY (TSC) was announced on July 27, 2005. Jointly owned by Virgin and Scaled Composites, the aerospace production company built launch aircraft and spaceships for spaceline operators. Virgin Galactic placed the first order for five second-generation SpaceShipTwo (SS2) passenger spaceplanes and two WhiteKnightTwo (WK2) carrier aircraft (see pp.72-73 and 76-77).

Having FAITH
The Spaceship Company's (TSC) operating headquarters are located at the Final Assembly, Integration, and Test Hangar (FAITH) at Mojave Air and Space Port, California. The vast building can accommodate the production of one WK2 and at least two SS2 vehicles in parallel.

IN NUMBERS:

NUMBER OF BASES: 3

NUMBER OF EMPLOYEES: 145

TOTAL FLOOR SPACE OF TSC OPERATIONS:
130,000 sq ft (12,100 sq m)

FAITH SPACECRAFT ASSEMBLY SPACE:
68,000 sq ft (6,320 sq m)

WIDTH OF FAITH HANGAR DOOR: 230ft (70m)

GROUNDBREAKING EVENT
On November 9, 2010, TSC broke ground for its new home on Spaceship Landing Way. Director of Operations Enrico Palermo addresses the gathering of VIPs.

HANGAR DOORS
The hangar's 230ft (70m) doors are wide enough to accommodate WK2s and SS2s large wingspan.

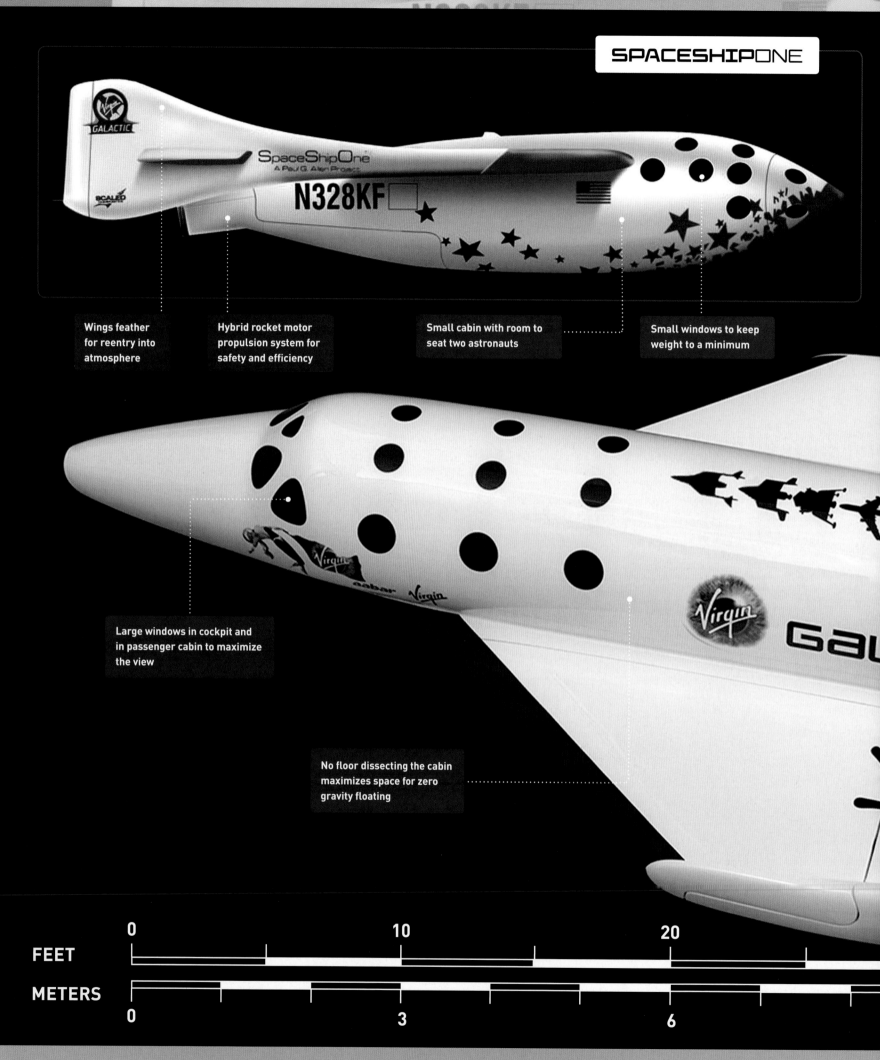

Wings feather for reentry into atmosphere

Hybrid rocket motor propulsion system for safety and efficiency

Small cabin with room to seat two astronauts

Small windows to keep weight to a minimum

Large windows in cockpit and in passenger cabin to maximize the view

No floor dissecting the cabin maximizes space for zero gravity floating

FEET

0   10   20

METERS

0   3   6

The same but different
SpaceShipOne was built to prove that getting into space could be done safely and cheaply. SpaceShipTwo is a dedicated passenger spacecraft, designed with the wants and needs of paying customers in mind. While some of the technology could be scaled up, issues such as maximizing cabin space and windows needed a complete rethink.

SCALING UP

USING THE SAME BASIC TECHNOLOGY as SpaceShipOne, Virgin Galactic's SpaceShipTwo does everything on a bigger scale. It is approximately twice as large as its predecessor, and is designed to carry six astronauts and two pilots. With similar dimensions to a Falcon 900 executive jet, every seat in the cabin is a window seat, with one side window and one overhead window for each astronaut.

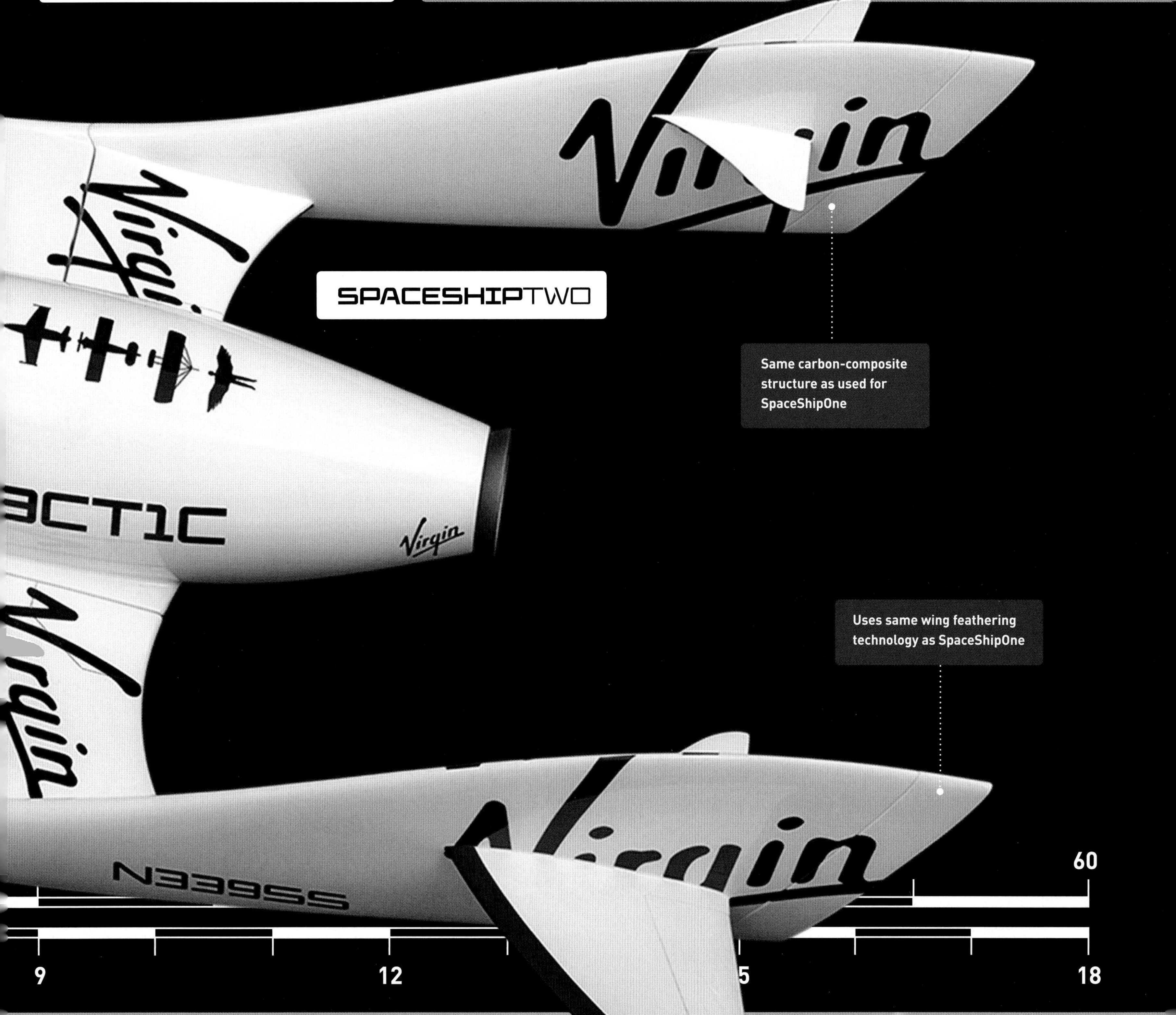
SPACESHIPTWO
Same carbon-composite structure as used for SpaceShipOne
Uses same wing feathering technology as SpaceShipOne
N339SS
60
9
12
5
18

EVOLUTION OF WK2

THE STRIKING DESIGN OF WHITEKNIGHTTWO owes much to two remarkable predecessors—the Virgin Atlantic GlobalFlyer and the carrier aircraft for SpaceShipOne, White Knight. Each groundbreaking aircraft was designed by Burt Rutan and built by Scaled Composites. The airplanes are constructed entirely of carbon composite, making them strong yet lightweight enough to fly at high altitude.

WHITE KNIGHT

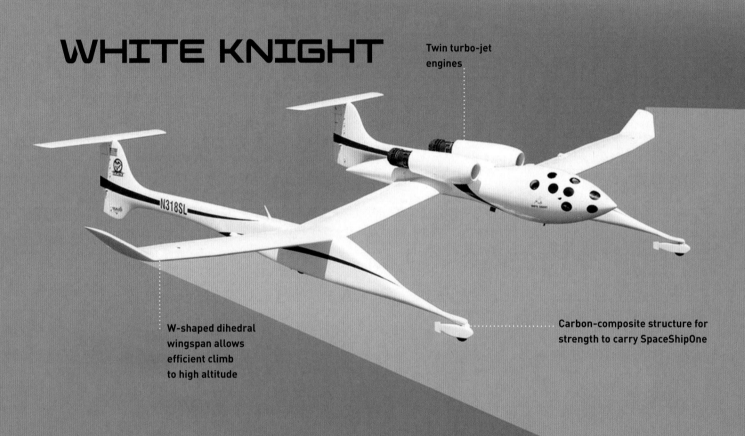

Twin turbo-jet engines

W-shaped dihedral wingspan allows efficient climb to high altitude

Carbon-composite structure for strength to carry SpaceShipOne

" ...THE APOGEE OF THE APPLICATION OF CARBON COMPOSITES TO AEROSPACE."

Sir Richard Branson, **Founder, Virgin Galactic**

VIRGIN ATLANTIC GLOBALFLYER

Carbon-composite structure
makes it lightweight and
fuel-efficient

Single pod
houses cockpit

Long wingspan for
high-altitude flight

Twin boom structure
increases stability

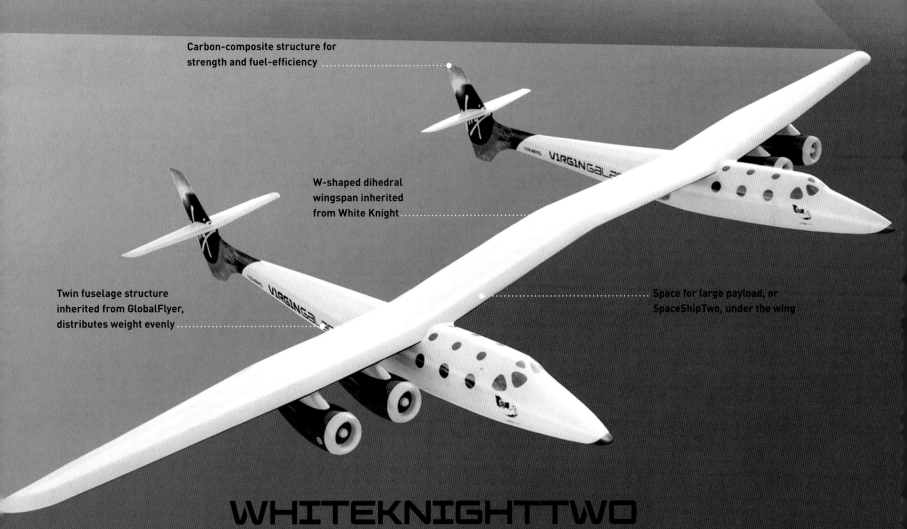

Carbon-composite structure for
strength and fuel-efficiency

W-shaped dihedral
wingspan inherited
from White Knight

Twin fuselage structure
inherited from GlobalFlyer,
distributes weight evenly

Space for large payload, or
SpaceShipTwo, under the wing

WHITEKNIGHTTWO

BUILDING WHITEKNIGHTTWO

CONCEALED WITHIN THE PRIVACY of its desert hangar, WhiteKnightTwo (WK2) takes shape. The vast hangar floor holds enormous armatures upon which the aircraft's long wingspan is formed from a single piece of carbon composite. The twin fuselages of WK2 are identical to that of SpaceShipTwo (SS2) and they are all manufactured in the same way, making the production process cheaper and simpler. The right fuselage on WK2 has a fully functioning cockpit, while the left one simply holds ballast with painted windows, for now.

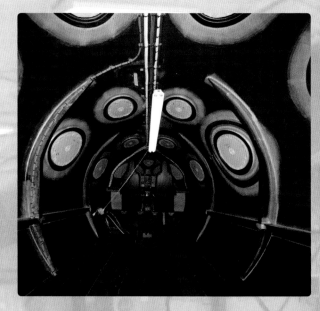

FUSELAGE
Each of WK2's twin fuselages is formed of a carbon-composite shell. The shells are shaped like gas canisters and narrow at one end. The window portholes are cut into the carbon.

UNDERSIDE

Virgin Galactic's logo was designed by Philippe Starck and adorns each wing. The blue used in the logo is based on the color of Richard Branson's eye (see p.66).

LIVERY

The DNA of Flight icons (see p.16) are displayed on the side of WK2's turbine. Here, the SS2 icon is depicted mounted between the wings of its mothership, placing WK2 at the forefront of aviation breakthroughs.

APPLICATION

A technician applies livery to the underside of WK2's wing. Each wing has flaps that can be raised to help WK2 drop as fast as SS2, meaning it can act as a chase plane and monitor SS2's descent.

WORK SPACE

The first WhiteKnightTwo was built in Scaled Composites' hangar. The carbon composite from which it is built lends itself to a curved design, meaning the aircraft has no sharp edges.

EVE IS BORN

WATCHED BY ONE HUNDRED future astronauts and the world's media, WhiteKnightTwo (WK2) left the secrecy of the hangar where it was built in Mojave, California on July 28, 2008. Christened *VMS Eve* in honor of Eve Branson, Richard's mother, WK2 represents some firsts of its own. It is the largest all carbon-composite aircraft ever built, and its 140ft (43m) wingspan is the longest flight component ever manufactured from the material. The unveiling of the mothership marked a significant milestone on Virgin Galactic's path to the first commercial spaceflight.

TS
THE SPACESHIP

CHRISTENING EVE
Mojave, California
• July 28, 2008

Richard Branson and his mother, Eve, toast the newly constructed *VMS Eve* at a naming ceremony. The event was attended by Virgin Galactic founder astronauts, who had already signed up to be carried away by the aircraft.

VMSEVE

"...IT REPRESENTS A FIRST AND A NEW BEGINNING. THE CHANCE... TO SEE OUR WORLD IN A COMPLETELY NEW LIGHT."

Sir Richard Branson, **Founder, Virgin Galactic**

WHITEKNIGHTTWO THE MOTHERSHIP

SLEEK BUT SUPREMELY STRONG, thanks to its carbon-composite form, WhiteKnightTwo (WK2) is capable of lifting heavy payloads to high altitude. Remarkably, for its size, WK2 can match SpaceShipTwo's (SS2) descent speed—making it a perfect simulator for the final stage of the mission.

TECHNICAL FILE:

WINGSPAN: 140ft (43m)

LENGTH: 78ft (24m)

TAIL HEIGHT: 25ft (8m)

CREW: 2-3

POWERPLANT: 4 × Pratt & Whitney Canada PW308 turbofan engines, 6,900lbf (30.69kN) thrust each

CAPACITY: payload 37,000lb (17,000kg) to 50,000ft (15km)

Keeping stable

With its twin cabins, twin tails, and long w-shaped wing, WhiteKnightTwo's structure makes it incredibly stable. The cabins are identical in structure, but the aircraft is piloted only from the right-hand cockpit.

READY TO ROLL

The first WhiteKnightTwo, christened *VMS Eve*, leaves the hangar at Scaled Composites. Future WhiteKnightTwos will be built by The Spaceship Company (see pp.70-71).

TURBOFAN

WhiteKnightTwo uses four highly efficient turbofan jet engines. They provide sufficient thrust for WK2 to carry SS2 to high altitude.

"THIS IS A PROGRAM THAT CAN'T HAVE A **HARD END DATE** AS **SAFETY** IS THE NUMBER ONE PRIORITY."

Stephen Attenborough, Commercial Director, Virgin Galactic, on the evolving test flight programme

POISED FOR TAKEOFF
Mojave, California
• December 21, 2008, *8:17am*
After the first WhiteKnightTwo was completed and successfully passed a rigorous round of ground tests, Virgin Galatic's test flight program began in earnest. Risks are kept to a minimum by testing exhaustively at each stage, and being confident of success before moving the program on.

TEST FLIGHTS

AN EXHAUSTIVE TEST PROGRAM for WhiteKnightTwo (WK2) and SpaceShipTwo (SS2) began with ground tests of the carrier aircraft. The process is at the heart of Virgin Galactic's commitment to safety. Once the ground tests were complete, WK2 took to the sky for its solo maiden flight before carrying SS2 as payload. The tests then progressed to allow SS2 to glide independently before, finally, making a powered flight.

FLIGHT PROFILE:

FLIGHT DURATION: I hour

TAKEOFF TIME: 8:17am

LANDING TIME: 9:17am

FLIGHT READY

After completing a comprehensive program of ground tests, *VMS Eve* is prepared for its maiden flight. The crew was confident that the aircraft would perform perfectly.

" ...WE HAVE A **SOLID PERFORMER** THAT **FLIES BEAUTIFULLY** AND WILL **SOON BE MOVING** ON TO ITS TASKS OF **LAUNCHING SPACESHIPS. "**

Sir Richard Branson, Founder, Virgin Galactic

"IT ALL WENT WELL... ALL THE BIG THINGS WORKED WELL."

Dick Rutan, aviator, on witnessing the test flight

VMS EVE MAIDEN FLIGHT
Mojave, California

• *December 21, 2008*

At 8:17am, *VMS Eve* took off from Mojave Air and Space Port. The test flight lasted an hour, during which the plane rose to 16,000ft (4.8km)—4,000ft (1.2km) higher than the intended maximum altitude for the flight, a testament to the crew's confidence. It made a flawless landing.

AT THE CONTROLS
INSIDE THE COCKPIT

THE COCKPITS OF WHITEKNIGHTTWO and SpaceShipTwo (SS2) are identical. They both accommodate two pilots with two sets of controls upfront. Space is tight and the control sticks are only around 8in (20cm) apart. Each window is 21in (53cm) across to give the pilots maximum visibility for landing.

WILL WHITEHORN IN THE SS2 COCKPIT
Former Virgin Galactic President,
Will Whitehorn, stands inside
the SpaceShipTwo cockpit.

ALL ABOARD THE MOTHERSHIP
The mothership has an additional seat
behind the two pilots, in which Sir Richard
Branson is sitting during this test flight.

BUILDING SPACESHIPTWO

BUILDING SPACESHIPS TAKES UP a lot of room. The team of engineers and fabricators working on SpaceShipTwo used enormous frames and molds to construct the shell of the ship in the vast Scaled Composites' hangar. The facilities are large enough to accommodate the molding and assembly stages at the same site.

STRONG SHELL

The carbon-composite shell of SpaceShipTwo's fuselage rests on an armature in the Scaled Composites hangar. The shell is molded before the components are assembled on the frame.

CARBON-COMPOSITE COMPONENTS for SpaceShipTwo (SS2) are molded into shape and put together on huge armatures. The many windows lining the cabin and cockpit are cut directly into the carbon-composite shell. The development of SS2 from the design and technology of SpaceShipOne (SS1) has taken many years—refinements and improvements will be an ongoing project.

"THERE ARE UNIQUE, NEW IDEAS SCATTERED THROUGHOUT THE SPACESHIP."

Burt Rutan, Founder, Scaled Composites

WOVEN CARBON
Strands of carbon fiber are woven together to form the casing that will store the fuel for SS2's rocket. It is replaced after each flight.

FUEL TANK
Two technicians work on the oxidizer tank, while it is suspended on to a frame. The tank will hold liquid nitrous oxide (laughing gas) and can be reused.

IN THE TANK
One technician climbs inside the large oxidizer tank to check the aluminum-lined interior, while another watches from the nozzle end.

TEAMWORK
The team prepares a giant mold to receive the carbon composite that forms the fuselages of SS2 and WK2. The carbon composite is layered for maximum strength and resistance. Once cured in the mold, it becomes incredibly stiff.

ROOM WITH A VIEW
The cabin is dotted with windows to ensure that each passenger has a view of Earth's curvature and the black of space. The hull is formed of a double layer of carbon with a honeycomb-like layer between them.

FINE FORM
The carbon-composite shells for both SS2 and WK2 fuselages are formed in giant molds, such as the one being worked on above. The carbon is applied to the mold and then allowed to cure in that shape.

INSIDE SPACESHIPTWO

EVERYTHING IN THE CABIN OF SPACESHIPTWO is made to provide the ultimate experience for the astronauts. The colors in this early concept design by Philippe Starck are simple, such as white and silver, to avoid distracting from the views outside. The final design for the spaceship will be revealed in 2014.

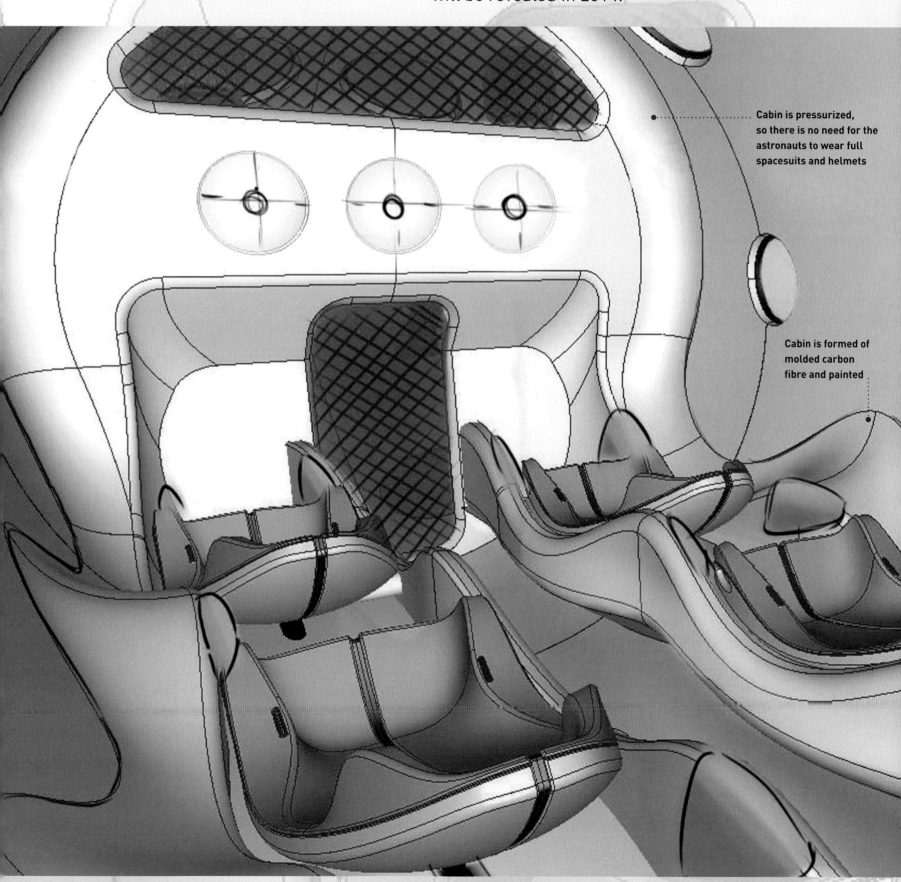

Cabin is pressurized, so there is no need for the astronauts to wear full spacesuits and helmets

Cabin is formed of molded carbon fibre and painted

"EVERYWHERE, I'M LOOKING TO REACH **ELEGANCE** AND INTELLIGENCE."

Philippe Starck, designer of concept interior for SpaceShipTwo

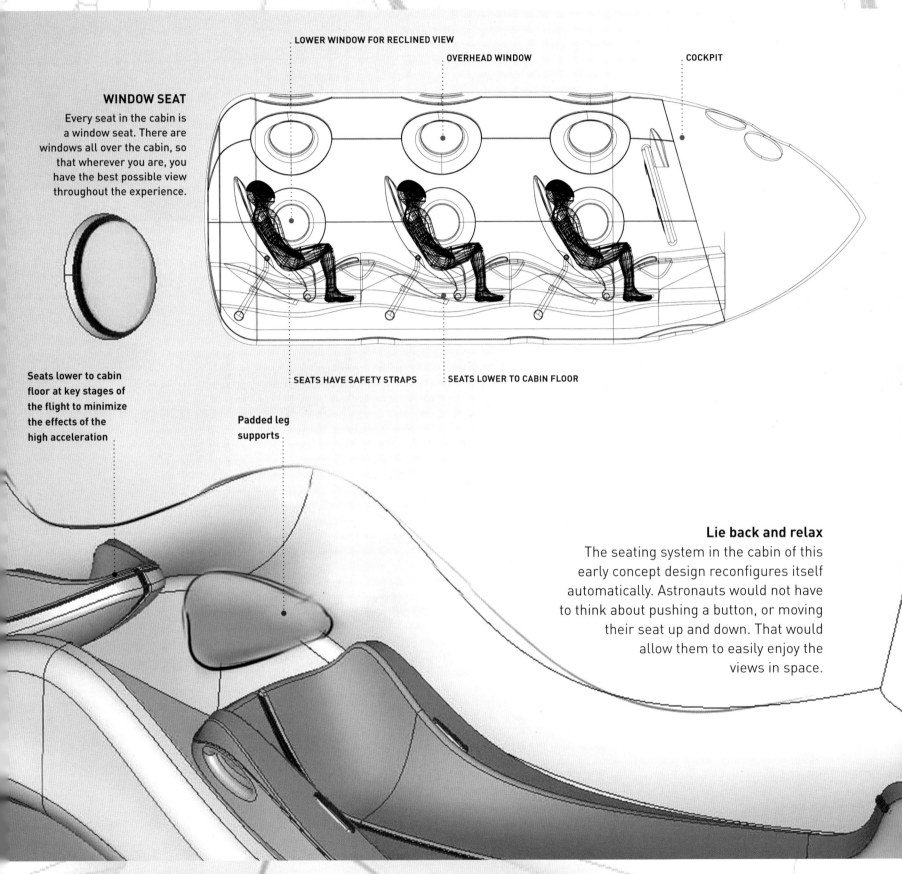

LOWER WINDOW FOR RECLINED VIEW

OVERHEAD WINDOW

COCKPIT

WINDOW SEAT

Every seat in the cabin is a window seat. There are windows all over the cabin, so that wherever you are, you have the best possible view throughout the experience.

SEATS HAVE SAFETY STRAPS

SEATS LOWER TO CABIN FLOOR

Seats lower to cabin floor at key stages of the flight to minimize the effects of the high acceleration

Padded leg supports

Lie back and relax

The seating system in the cabin of this early concept design reconfigures itself automatically. Astronauts would not have to think about pushing a button, or moving their seat up and down. That would allow them to easily enjoy the views in space.

SPACESHIPTWO
MEET VSS ENTERPRISE

VSS ENTERPRISE IS THE FIRST of five SpaceShipTwo (SS2) aircraft to be produced for commercial flights. The passenger spaceplane's name is a nod to the *USS Enterprise*, a fictional spacecraft featured in the TV series, *Star Trek*.

TECHNICAL FILE:

CREW: 2

CAPACITY: 6 passengers

LENGTH: 60ft (18.3m)

WINGSPAN: 27ft (8.3m)

TAIL HEIGHT: 18ft (5.5m)

LOADED WEIGHT: 21,428lb (9,740kg)

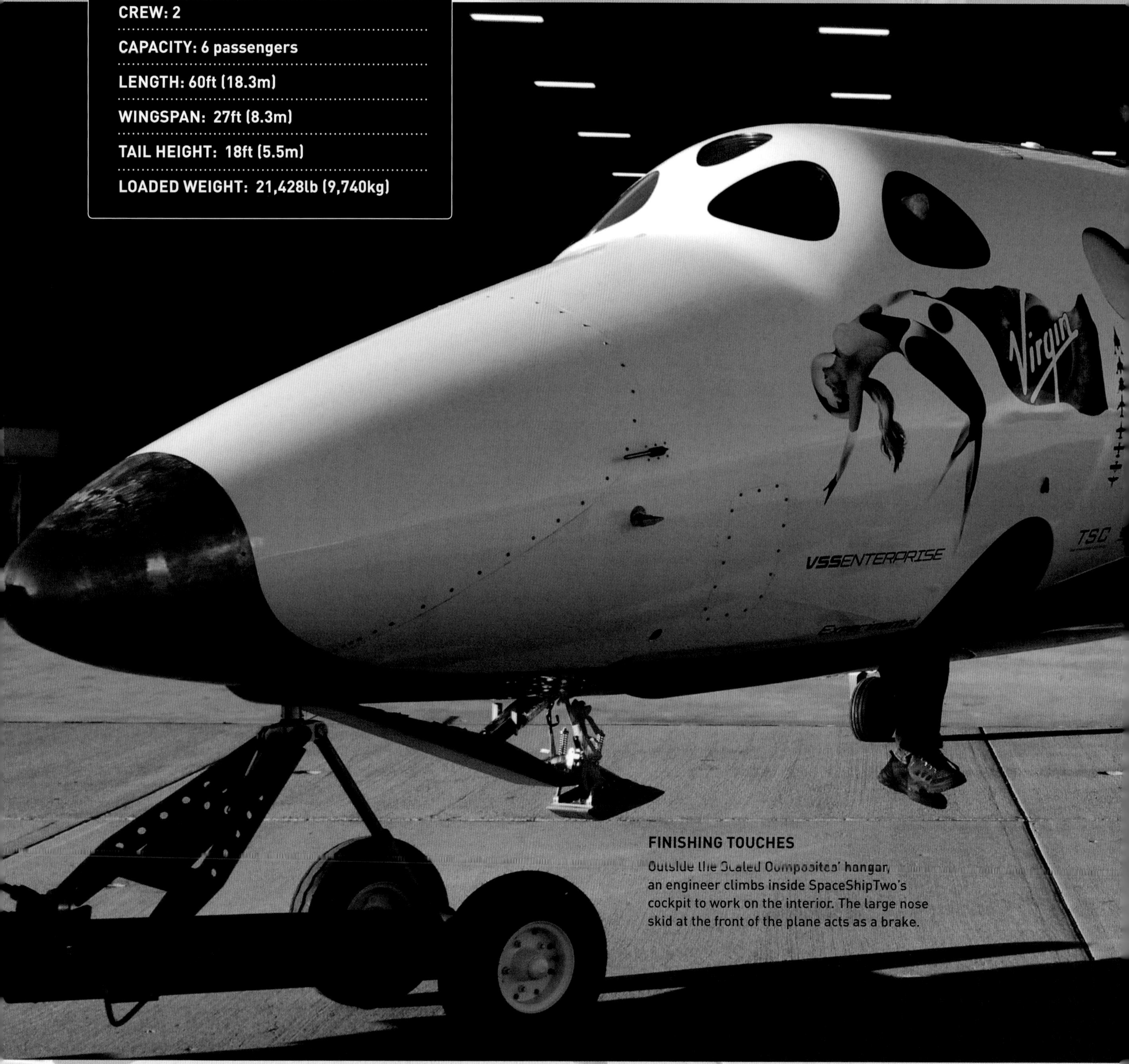

FINISHING TOUCHES

Outside the Scaled Composites' hangar, an engineer climbs inside SpaceShipTwo's cockpit to work on the interior. The large nose skid at the front of the plane acts as a brake.

ROCKET CONE

SpaceShipTwo's hybrid rocket motor ends in a conical nozzle. The team who built it signed the casing. SpaceShipTwo's rocket uses the same means of propulsion as SpaceShipOne and burns benign fuel with an oxidizer (see p.48). The rocket can be controlled by the pilot and can be shut down, if necessary, during the boost stage of the flight.

STABILIZERS

The angled tips of SpaceShipTwo's wings tilt upwards to stabilize the spaceship during its descent.

SPACESHIPTWO & WHITEKNIGHTTWO

AS THE WORLD'S FIRST COMMERCIAL spaceline, Virgin Galactic operates two remarkable vehicles. WhiteKnightTwo (WK2) is the largest carbon-composite carrier vehicle in service. Refining the technology of SpaceShipOne (SS1), SpaceShipTwo (SS2) is essentially a glider with the capability to reach speeds of Mach 3.5.

AS WELL AS BEING A MACH 3.5 SPACESHIP, SS2 IS A REMARKABLE HIGH-ALTITUDE GLIDER.

TECHNICAL FILE:

CABIN DIAMETER:
SS2 and WK2: 7.5ft (2.3m)

PROPULSION:
SS2: 1 controllable hybrid rocket motor
WK2: 2 turbofan jet engines

SS2 ROCKET BURN TIME:
60 seconds

GEARS:
SS2: tricycle configuration; 2-wheeled main gears; 1 nose skid; retractable
WK2: quadricycle configuration; retractable

140FT/43M

25FT/8M

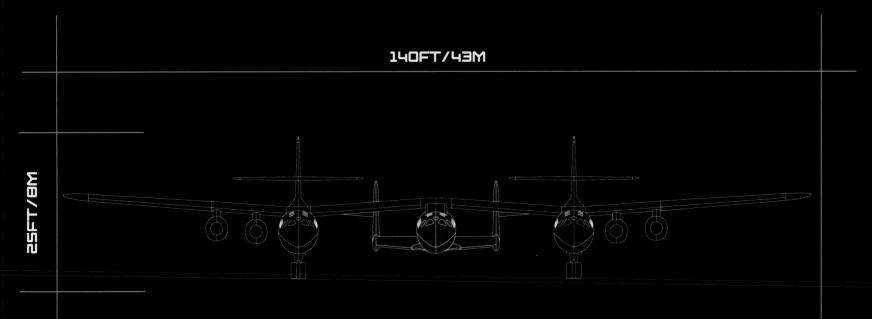

"...AN INCREDIBLE VEHICLE THAT IS GOING TO OPEN UP SPACE TO MORE PEOPLE THAN EVER BEFORE"

Dave Mackay, Chief Pilot, Virgin Galactic, on SpaceShipTwo

Room for SS2
With almost 50ft (15m) separating the twin fuselages of WK2, the 42-ft (13-m) wingspan of SS2 sits comfortably under the apex of the mothership's w-shaped wing.

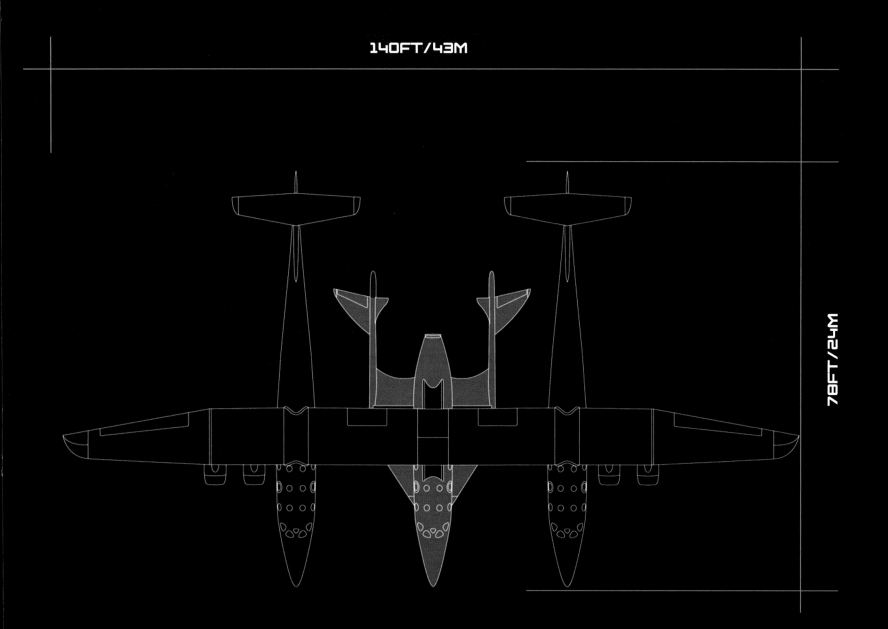

140FT/43M

78FT/24M

DAVE MACKAY
Virgin Galactic's Chief Pilot, who will be at the controls for the first commercial flight

ADAM WELLS
Head of Design, responsible for interiors of Spaceport America and SS2

STEPHEN ATTENBOROUGH
Commercial Director and Virgin Galactic's first full-time employee

VIRGIN TERRITORY
Mojave, California

Virgin Galactic employees gathered in the hangar at Mojave to see SpaceShipTwo (SS2) and WhiteKnightTwo (WK2). Team members uprooted their lives and families to move to the remote area when they joined the company. The people on the ground making it all happen are pioneers— breaking new ground in the fledgling commercial spaceflight industry.

EVE BRANSON
Mother of Sir Richard Branson—*VMS Eve* was named in her honor

ENRICO PALERMO
Vice President of Operations, The Spaceship Company— oversaw development of SS2

UNVEILING ENTERPRISE

DESPITE ADVERSE WEATHER CONDITIONS, more than 800 press, future astronauts, and VIPs gathered to catch the first glimpse of SpaceShipTwo leaving the Scaled Composites' hangar in Mojave, California. The world's first commercial spaceship was carried down the runway by the mothership to a light and music display. Richard Branson's daughter, Holly, christened the ship *VSS Enterprise*.

WEATHERING THE STORM
Mojave
• *December 7, 2009*
Pushing through the gale-force winds, snow, and stormy weather, SpaceShipTwo emerges from the hangar on the wing of *VMS Eve* amid a spectacular light display.

"THIS IS A MOMENTOUS DAY FOR THE SCALED AND VIRGIN TEAMS."

Burt Rutan, Founder, Scaled Composites

CAPTIVE CARRY
Mojave, California
• *March 22, 2010*

VSS Enterprise remained firmly attached to carrier aircraft *VMS Eve* during its maiden flight. From takeoff at 7:05am to landing almost three hours later, everything went to plan. The craft reached a peak altitude of 45,000ft (13.7km), paving the way for the next stage of testing—independent glide flights.

FLYING FREE

FOLLOWING FOUR CAPTIVE CARRY TESTS, SpaceShipTwo (SS2) was ready for its true maiden flight. The spaceplane was released from WhiteKnightTwo at more than 40,000ft (12.2km) and glided back to the runway on October 10, 2010. The main objective of the flight was to test the release mechanism. On reaching the target altitude, SS2 dropped away cleanly. The pilots flew freely before making a practice approach to the runway and gliding into land.

IN NUMBERS:

FLIGHT TIME: 13 min

ALTITUDE AT RELEASE: 46,000ft (14km)

FIRST RELEASE

With a gentle bump, SpaceShipTwo drops away from the mothership. The air-launch mechanism works in the same way as that of White Knight and SpaceShipOne.

First to fly

Piloting experimental aircraft takes particular skill, courage, and understanding. Each test flight has set objectives that require the aircraft to be flown in a specific way. The Scaled Composite test pilots for SS2's first glide flight were Mike Alsbury and Pete Siebold.

Co-pilot: Mike Alsbury
Project engineer and test pilot at Scaled Composites, Mike Alsbury has more than 15 years of flight experience.

Pilot: Pete Siebold
Test pilot Pete Siebold has more than 17 years of flight experience and worked on the Tier One project with Burt Rutan.

"AFTER WE LANDED, I LOOKED OVER TO MIKE AND SAID, 'CAN WE DO THAT AGAIN?'"

Pete Siebold, Test Pilot, Scaled Composites

FIRST FREE FLIGHT
Mojave, California

• *October 10, 2010*

VSS Enterprise was released cleanly from the carrier aircraft *VMS Eve* at around 46,000ft (14km) during its maiden flight. The pilots, Pete Siebold and Mike Alsbury performed selected test maneuvers reaching up to 2gs. The free flight lasted 13 minutes, ending with a perfectly targeted glide and landing. The spaceplane performed superbly.

FEATHERING FOR SAFE REENTRY

AT THE FLICK OF A SWITCH, SpaceShipTwo (SS2) turns into a shuttlecock. Burt Rutan's ingenious, yet simple design was first developed for SpaceShipOne. It provided a solution to the hazards of reentry from the vacuum of space to the dense atmosphere. SS2's wings pivot up, meaning the aerodynamics of the raised section slow and control the spaceship by creating a shuttlecock-like drag.

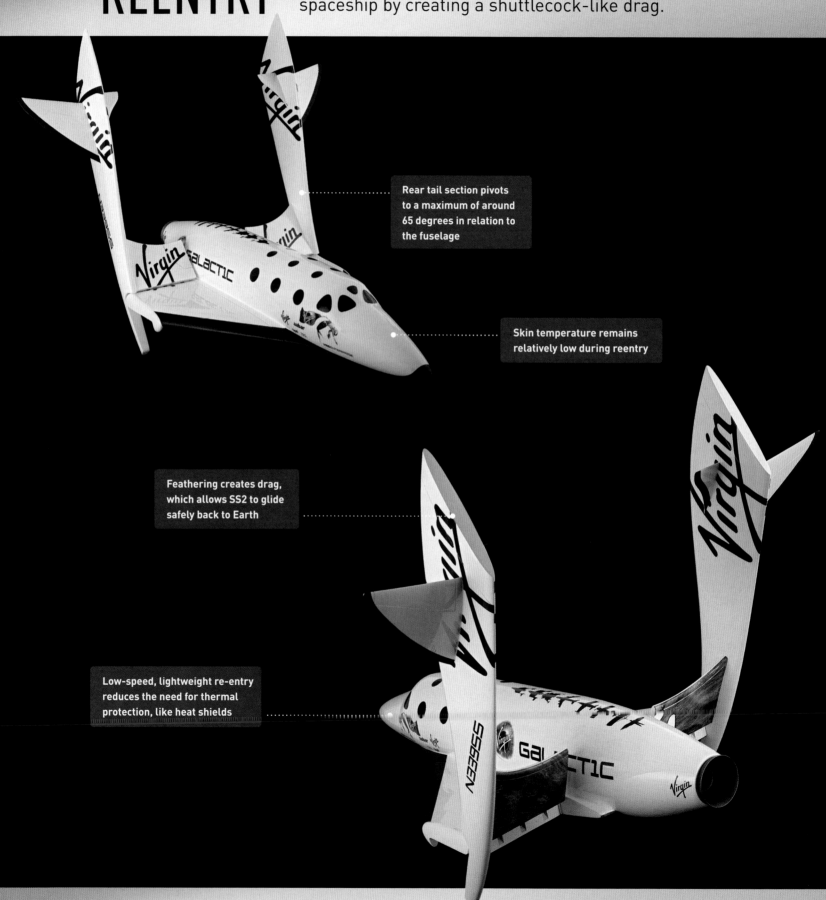

Rear tail section pivots to a maximum of around 65 degrees in relation to the fuselage

Skin temperature remains relatively low during reentry

Feathering creates drag, which allows SS2 to glide safely back to Earth

Low-speed, lightweight re-entry reduces the need for thermal protection, like heat shields

Pivoting wings
Once out of Earth's atmosphere the tail section of SS2 can be rotated up to around 65 degrees. This configuration allows for automatic control of the spacecraft's position. The fuselage stays parallel to the atmosphere, keeping SS2 stable. The ship's low weight and high drag means that the exterior does not overheat during reentry.

Rudders are activated by pilot's pedals

Elevons activated by pilot's center stick for vertical pitch and roll control

Rocket motor system

Feather mechanism is activated and locked hydraulically by pilot

Roll thrusters control pitch of spaceship

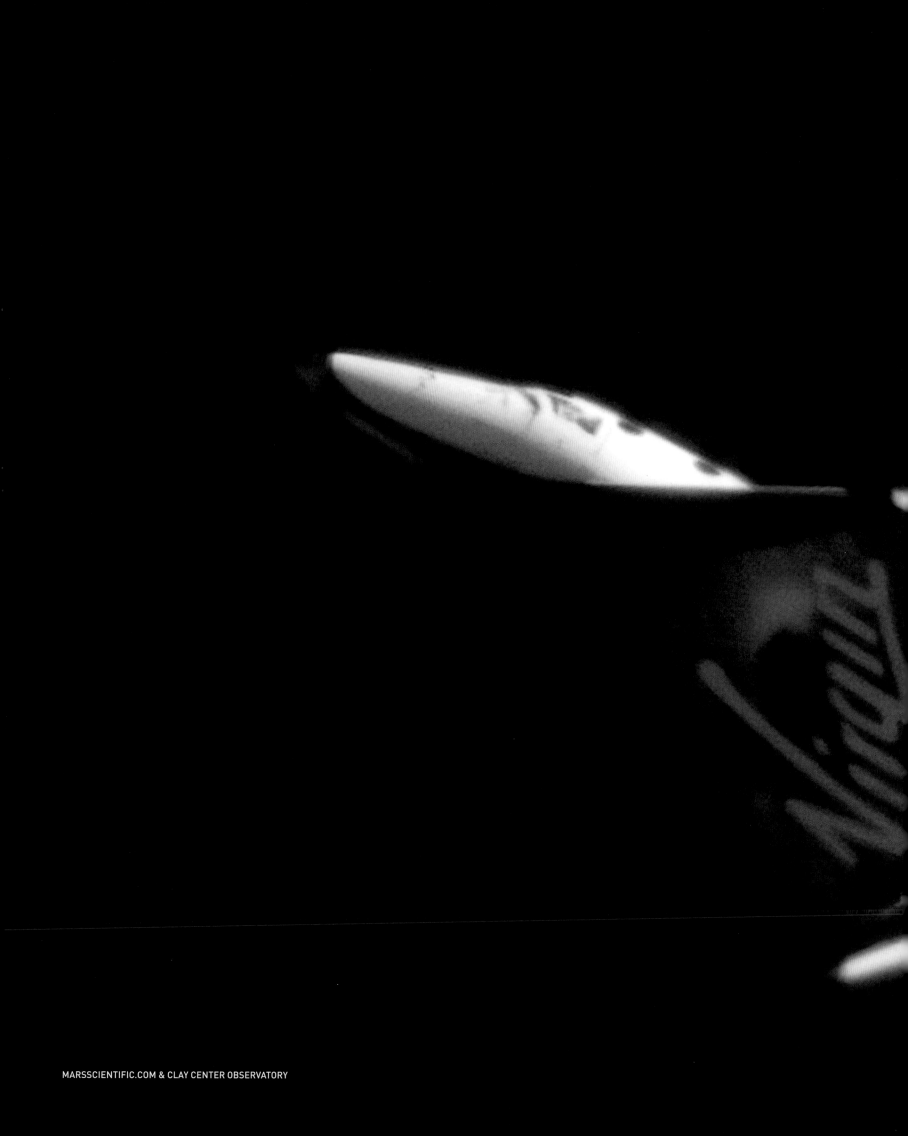

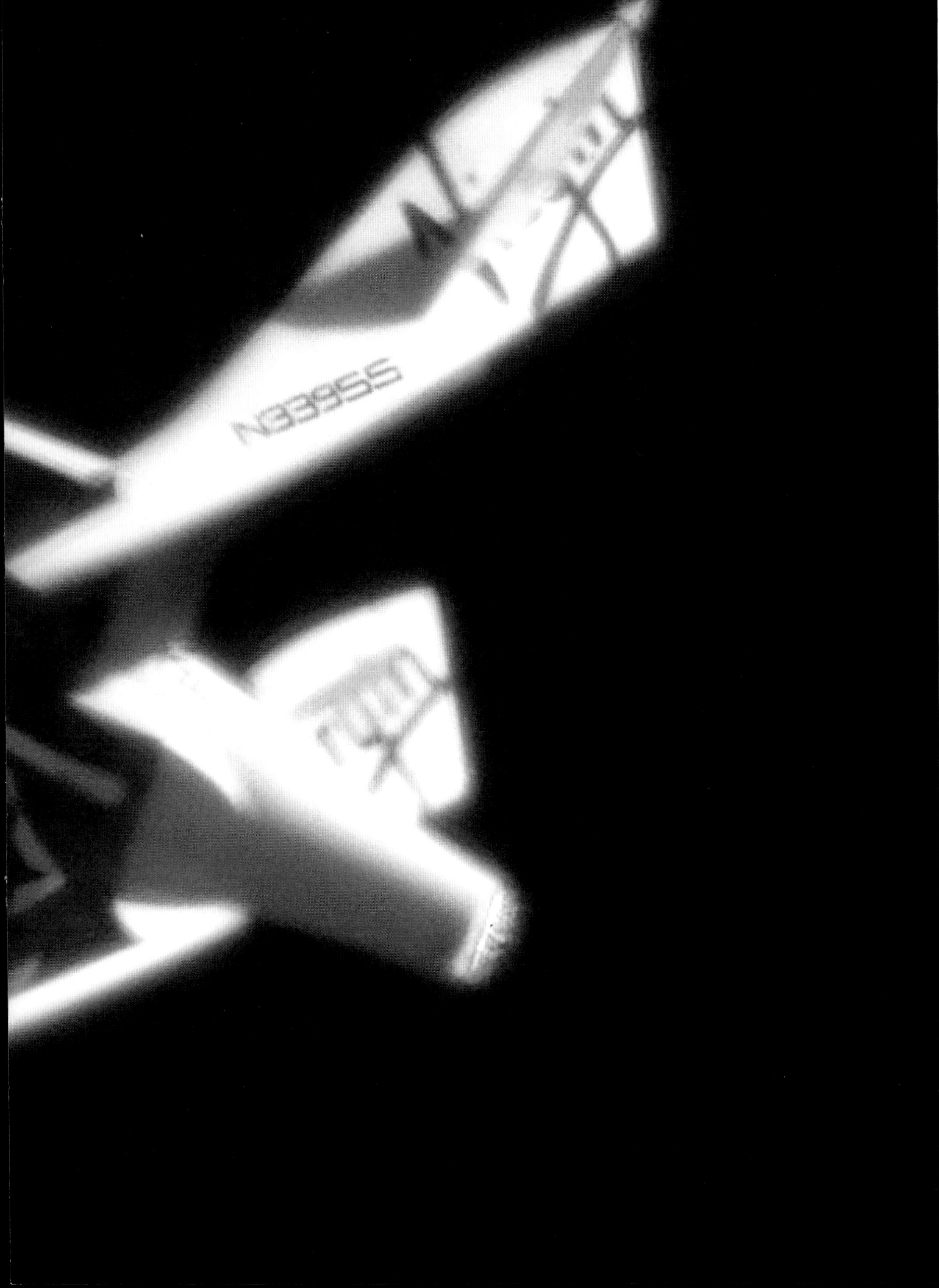

"THIS MORNING'S FLIGHT WAS A TEST PILOT'S DREAM."

Pete Siebold, Test Pilot, Scaled Composites

FIRST FEATHERED FLIGHT
Mojave, California

• *May 4, 2011*

On its seventh solo flight, *VSS Enterprise* used its most remarkable safety feature for the first time. After release from *VMS Eve* at 46,000ft (14km), *VSS Enterprise* went into a stable glide before deploying its unique feathered configuration. The tail section rotated up 65 degrees for just over a minute before reconfiguring to glide at 33,500ft (10km).

DOWN TO EARTH

ONCE BACK IN EARTH'S ATMOSPHERE, SpaceShipTwo (SS2) is designed to turn into an unpowered glider. Its early solo test flights focused on testing the release mechanism and how it performed in glide mode. With its wings pointing back, rather than in the upright "feathered" mode (see p.110), SS2 can glide in to land on a conventional runway. Its carbon-composite structure makes it strong and light, while the wing shape and strakes provide stability.

Landing sequence
The final stage of SpaceShipTwo's flight is completely unpowered. The rocket is spent and the wings are folded back for a gentle glide down to Earth.

HOME STRAIGHT
SpaceShipTwo glides back to the runway on its first solo flight. Pilot Pete Siebold commented that it was "a joy to fly," considering that it was built to reach supersonic speeds.

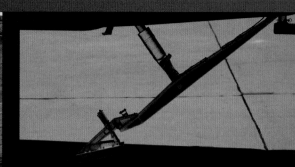

1 **WINGS BACK**
The wings tilt back out of feathered mode, and the body of SS2 returns to a horizontal alignment.

2 **WHEELS DOWN**
As the spaceplane approaches the runway, two wheels are lowered from the fuselage to meet the tarmac.

3 **NOSE SKID**
A large nose skid under the front of SS2 acts as a brake after landing. It is fitted with an abrasive shoe that slows the plane down.

"...I WATCHED THE WORLD'S FIRST MANNED COMMERCIAL SPACESHIP LANDING ON THE RUNWAY... IT WAS A GREAT MOMENT."

Sir Richard Branson, Founder, Virgin Galactic

Wings set into horizontal position to make SpaceShipTwo an aerodynamic, high-altitude glider

SPACEPORT AMERICA

DEEP IN THE DESERT LANDSCAPE OF NEW MEXICO, the world's first commercial spaceport began to take shape in 2006. The state was eager to become involved in the space tourism business, and voters approved a tax to part-fund the build, amounting to $40 million of public money. A suitable site was located in the Jornada del Muerto desert basin, and British architects Foster & Partners were commissioned to design the building.

IN NUMBERS:

COST OF BUILDING SPACEPORT:
$209 million

LENGTH OF LEASE VIRGIN SIGNED TO HAVE SPACEPORT AS THEIR HQ IN 2008:
20 years

SIZE OF SPACEPORT AMERICA:
27 sq miles (70 sq km)

"WE ARE CELEBRATING...THE WORLD'S FIRST PURPOSE-BUILT, COMMERCIAL SPACEPORT."

Bill Richardson, Governor of New Mexico

METAL FRAMEWORK

The early stages of construction saw the supportive concrete base erected. A metal framework was then constructed over and around it.

SKY LIGHT

The roof of Spaceport America is covered in skylights. By maximizing the use of natural light, energy use and costs are kept to a minimum.

GATEWAY TO SPACE

Visitors enter Spaceport America via a deep channel cut into the landscape. The main structure of the building envelopes the entrance.

MAKING PROGRESS

By 2010, construction was well underway and the exterior was taking shape. The paneled roof was substantially in place, and SpaceShipTwo flew over to see its future home.

DESERT HUB

LYING LOW WITHIN THE LANDSCAPE, the spaceport's organic form appears to be no more than a gentle rise in the hilly terrain. Built using local material and regional construction techniques, the building is both sustainable and sensitive to its surroundings. Dug deep into the land itself, Spaceport America uses the thermal mass as a buffer against the extreme climate.

FITTING IN
The spaceport was designed with its desert surroundings in mind. It is insulated by the earth it lies in, lit by the ample New Mexico sunlight, and ventilated by the westerly winds.

"THIS IS GOING TO BE A CRUCIAL HUB FOR THE COMMERCIAL EXPANSION OF SPACE."

Will Whitehorn, former President, Virgin Galactic

GROUND CONTROL

FOR SUCCESSFUL FLIGHTS, both in a test program and in commercial operations, a dedicated control team on the ground is vital. Communicating with pilots during each stage of a mission—from taxiing through liftoff to landing—the control room continually assesses the situation. The team's primary task is to ensure safety, and to provide information and support for the pilots.

SPACE TAXI

WhiteKnightTwo and SpaceShipTwo taxi to the runway at Mojave Air and Space Port prior to a test flight. They are followed by a ground support vehicle and a chase plane that will monitor the flight.

Flight support
It is the responsibility of the control team to monitor traffic in the air and on the ground. Each flight is given a time slot for takeoff and landing to ensure safety. Ground crews also carry out checks on the aircraft before sending them off, and they are on standby for returning flights.

View from above: SpaceShipTwo touches down on the runway in Mojave, California, dwarfed by the jet planes that are parked along the tarmac.

Ground crew: Flight engineers carry out last-minute checks on both vehicles as they wait on the tarmac at Mojave before a test flight.

COMING HOME

VIRGIN GALACTIC MADE Spaceport America in New Mexico its home in 2008. The terminal hangar facility will house up to two motherships and five SpaceShipTwos, and will serve as the operating hub. There will also be extensive astronaut preparation facilities and mission control. In 2011, a combined terminal hangar at Spaceport America was dedicated as the Virgin Galactic Gateway to Space.

DESERT BASE

The rear of Spaceport America looks out on to the runway. From above, the shape of the spaceport evokes the Virgin Galactic iris logo, with the building itself forming the pupil and the surrounding tarmac as the iris.

SPACIOUS SURROUNDINGS

VMS Eve and *VSS Enterprise* try the cavernous Spaceport America hangar for size. It will one day house the Virgin Galactic fleet.

ON REFLECTION

The mirrored windows reflect the desert landscape, let in natural light, and provide impressive viewing points.

INTERIOR

The vast, high-ceilinged spaces inside the terminal building are designed to be light, airy, and futuristic.

BALCONY

An elevated viewing gallery runs around the outside of the terminal. Visitors and future astronauts can enjoy views of the surrounding desert and watch takeoffs and landings.

KEYS TO A NEW DAWN

More than 150 future astronauts gathered at the spaceport in October 2011 for the "Keys to a New Dawn" event, which christened the Virgin Galactic Gateway to Space. Acrobatic dancers scaled the terminal building.

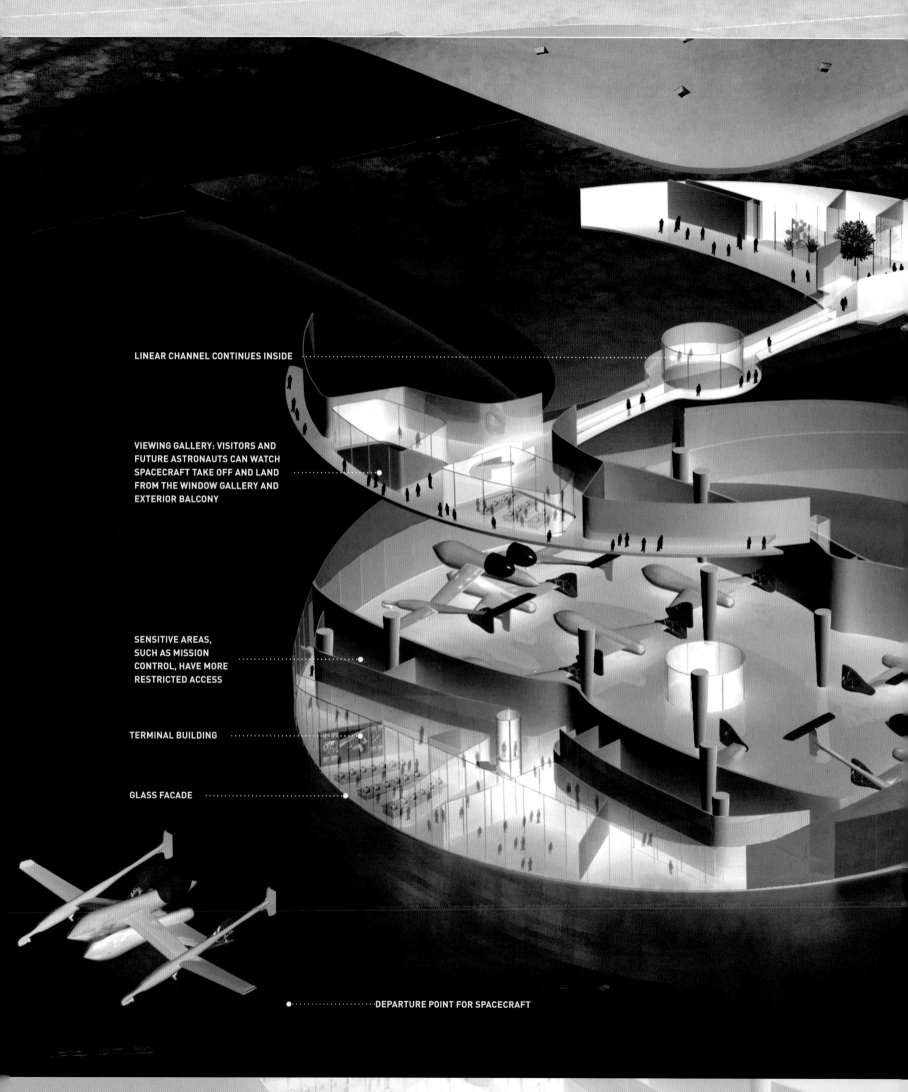

LINEAR CHANNEL CONTINUES INSIDE

VIEWING GALLERY: VISITORS AND
FUTURE ASTRONAUTS CAN WATCH
SPACECRAFT TAKE OFF AND LAND
FROM THE WINDOW GALLERY AND
EXTERIOR BALCONY

SENSITIVE AREAS,
SUCH AS MISSION
CONTROL, HAVE MORE
RESTRICTED ACCESS

TERMINAL BUILDING

GLASS FACADE

DEPARTURE POINT FOR SPACECRAFT

ENTRY CHANNEL ·············

AWE-INSPIRING
The spaceport is designed to reflect the mysetery and awe of space travel inside and out.

EXHIBITION SPACE: DOCUMENTS THE HISTORY OF THE REGION AND ITS SETTLERS, ALONGSIDE THE HISTORY OF SPACE EXPLORATION

SUPERHANGAR: WILL ONE DAY HOUSE THE VIRGIN GALACTIC FLEET OF WHITEKNIGHTTWOS AND SPACESHIPTWOS, ALONG WITH A SIMULATOR ROOM FOR TRAINING.

CLEAR VISION
This conceptual artwork shows the thinking behind the spaceport's design. Each layer is organized for ease of access according to its purpose.

RATIONAL SPACE

THIS CONCEPTUAL DESIGN of Spaceport America, which features a fleet of White Knights and SpaceShipOnes, is organized to run efficiently. A balance between accessibility and privacy is maintained throughout the building. The astronauts' areas and visitor spaces are fully integrated. The more sensitive zones, such as the control room, are visible, but have limited access.

SPACEWAY

CUTTING A LINE THROUGH THE DESERT SANDS,

the 2-mile (3-km) long spaceway stands ready to launch spacecraft into space. The hefty 42-in (1-m) thick spaceway can support almost every type of aircraft in the world today. It is finished with a 14in (36cm) layer of concrete, and it can accommodate returning launch vehicles, rocket boosters, and training vehicles.

Dedicating the runway

Governor of New Mexico, Bill Richardson, Sir Richard Branson, and approximately 30 future astronauts attended an event to dedicate the newly built spaceway at Spaceport America. WhiteKnightTwo completed a flyover carrying SpaceShipTwo, before coming in to land on the spaceway in front of the crowd.

PATH TO THE STARS

Overlooked by the glass-paneled facade of the spaceport, the spaceway stretches for 2 miles (3km) through the arid terrain of the New Mexico desert.

FUTURE ASTRONAUTS
Gathered at the rear of the spaceport, still under construction, a crowd of Virgin Galactic future astronauts view the spaceway. It will one day be the starting point of their journey into space.

MAKING IT HAPPEN
New Mexico's Governor, Bill Richardson, and Sir Richard Branson placed their handprints in clay at the naming ceremony for the runway, which is called "Governor Bill Richardson Spaceway." The Governor was key in pushing forward the spaceport project.

IN NUMBERS:

LENGTH: 2 miles (3km)

THICKNESS: 42in (1m)

COMPOSITION: 24in (61cm) of sub-grade, 4in (10cm) of asphalt, 14in (36cm) of concrete

ROCKET POWER

ONCE RELEASED FROM THE MOTHERSHIP, SpaceShipTwo (SS2) falls for a few seconds—and then the magic happens. The hybrid rocket motor ignites, and SS2 accelerates into a vertical climb. A burn-time of around 60 seconds is long enough to lift the ship into space. During the flight, the pilot can control the rocket—shutting it off mid-blast, if needed—and glide safely back down to the runway.

Fire test: New Mexico—3:00am
Utilizing similar technology to that of SpaceShipOne, the hybrid rocket motor burns solid rubber and nitrous oxide. Before testing the rocket in flight, it was necessary to complete extensive ground fire tests. Special frames were made to keep the rockets secure during the tests.

PURER POWER
At full blast, SpaceShipTwo's rockets will be able to achieve speeds of up to Mach 4—four times the speed of sound. Hot gases produced by the burn exit from the nozzle to produce thrust.

IN DEVELOPMENT
The hybrid rocket motor system was developed at Scaled Composites for SpaceShipOne. Comprehensive ground fire tests were carried out at the Mojave facility before the flight.

ROCKETMOTORTWO
SpaceShipTwo's hybrid rocket motor system, called RocketMotorTwo, undergoes a full duration test fire by the Sierra Nevada Corporation at their desert facility.

"TO BE BEHIND THE CONTROLS AND FLY IT AS THE ROCKET IGNITED IS SOMETHING I WILL NEVER FORGET."

Dave Mackay, Chief Pilot, Virgin Galactic

"AS PART OF OUR WONDERFUL, PIONEERING FUTURE ASTRONAUT COMMUNITY, YOUR PLACE IN HISTORY IS ASSURED."

Sir Richard Branson, Founder, Virgin Galactic

IN THEIR DNA
Mojave, California

• *September 25, 2013*

The largest ever gathering of future astronauts took place at Mojave Air and Space Port to mark the progress toward commercial flight. The event, which was called "Your Flight DNA," gave around 400 customers the chance to meet the pilots and view *VMS Eve* and *VSS Enterprise* outside The Spaceship Company's hangar.

"WE WANT TO GIVE OUR CUSTOMERS THE OPPORTUNITY TO MEET THE TEAM BEHIND THE DREAM..."

George Whitesides, CEO, Virgin Galactic

Going global

Virgin Galactic events take place all over the world. Future astronauts have visited Sir Richard Branson's island in the Caribbean, his game reserve in South Africa, and watched test flights in the Mojave desert in the USA.

Sweden: An event was held at the Ice Hotel, Kiruna, where it is hoped that Sweden will one day host a spaceport.

Necker Island: Sir Richard Branson invites groups of future astronauts to stay with him on his private island.

Exclusive: Richard Branson holds events for future astronauts on his private island (above) and reserve.

KEYS TO A NEW DAWN

Project Bandaloop, a team of vertical dancers, perform on the spaceport's facade at the "Keys to a New Dawn" event, which was held in 2011 to mark the completion of the building's exterior.

"IT IS LITERALLY **OUT OF THIS WORLD, AND THAT'S WHAT WE WERE AIMING AT CREATING."**

Sir Richard Branson, Founder, Virgin Galactic

ROYAL ROAD
Sierra County, New Mexico

To arrive at Spaceport America, visitors travel along an old desert trail called El Camino Real, meaning the "royal road". Set in a remote area of New Mexico, the Jornada del Muerto desert basin remained unpaved until 2012. Through the haze of the desert heat, the spaceport appears at once other-worldly, yet intrinsically part of this sinuous landscape.

ARRIVING AT SPACEPORT

DEEP IN THE NEW MEXICO DESERT, Spaceport America will be home to Virgin Galactic astronauts for a few days as they acclimate, and undergo checks and training before their flight. Entering the spaceport via a ramp cut into the landscape, visitors arrive at huge steel doors that open into a spacious astronaut lounge. At preflight briefings, astronauts meet the crew.

YOUR SHIP AWAITS

The spaceport's astronaut lounge sits over three stories. A wall of mirrored windows looks out onto the spaceway.

"LEARNING HOW TO MAKE THE MOST OF THE TIME... WILL FORM **AN IMPORTANT PART OF THE PREPARATION.**"

Stephen Attenborough, Commercial Director, Virgin Galactic

SPACEWAY

VIRGIN GALACTIC
GATEWAY TO SPACE

SPACEPORT
OPERATIONS CENTER

FUEL DEPOT

WATER STORAGE
FACILITY

SPACEPORT
CENTRAL

TOUR OF THE FACILITY
As part of the acclimatization program, future astronauts will get a tour of Spaceport America. The Virgin Galactic Gateway to Space terminal sits at the heart of the huge desert complex.

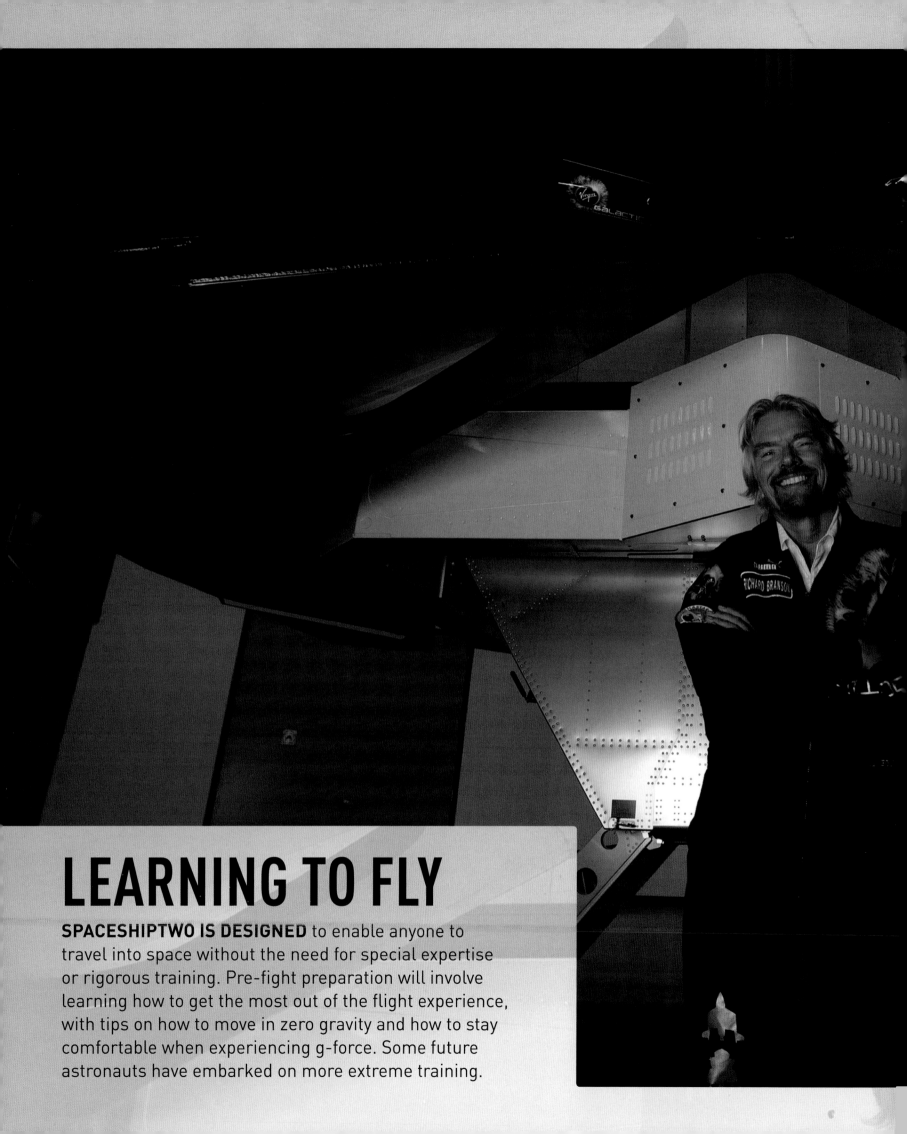

LEARNING TO FLY

SPACESHIPTWO IS DESIGNED to enable anyone to travel into space without the need for special expertise or rigorous training. Pre-fight preparation will involve learning how to get the most out of the flight experience, with tips on how to move in zero gravity and how to stay comfortable when experiencing g-force. Some future astronauts have embarked on more extreme training.

Getting flight fit

Richard Branson, his children, and a group of future astronauts visited the National Aerospace Training and Research Center (NASTAR) in Philadelphia to experience g-force in a centrifuge device:

Briefing: Richard Branson is told what to expect on entering the capsule. The physical effects are like going into space.

In a spin: Smiling through the pressure, Richard Branson experiences 6gs of force, produced by high acceleration.

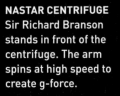

NASTAR CENTRIFUGE
Sir Richard Branson stands in front of the centrifuge. The arm spins at high speed to create g-force.

HOW HIGH WILL YOU FLY?

SPACESHIPTWO WILL CARRY Virgin Galactic customers through the layers of Earth's atmosphere, which blend into space as the spaceship travels higher. Those onboard will become astronauts and will be awarded a set of Astronaut Wings at a ceremony when they return to Spaceport America.

MAXIMUM PLANNED ALTITUDE, BEFORE UNPOWERED RETURN TO EARTH

"IT WAS A TEXTURE. THE BLACKNESS WAS SO INTENSE."

Charles Duke, Apollo 16 astronaut, on viewing space

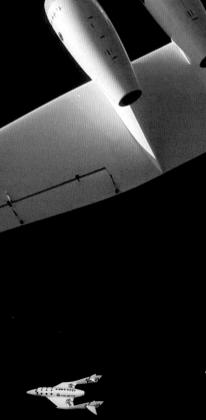

The mothership releases
SpaceShipTwo. The spaceship
then climbs to maximum altitude.

KITTED OUT

EACH VIRGIN GALACTIC ASTRONAUT will be fitted with a specially designed, personalized spacesuit for their flight. The suits will be softshell with soft-soled shoes. Full space helmets are not needed because the passenger cabin of SpaceShipTwo (SS2) is fully pressurized. Soft protective headgear may be worn, however, which may contain headphones and a microphone for communication.

SPACEMAN

Sir Richard Branson poses in a spacesuit on his private island, Necker, in the Caribbean. SS2 astronauts will wear different spacesuits.

Evolving style

The design for the Virgin Galactic spacesuit has evolved with the project. For safety, during the zero gravity experience, the suit needs to be soft and lightly padded with no hard edges, and it must allow freedom of

Soft shell: Richard Branson and his son, Sam, wear Virgin Galactic-branded suits for centrifuge training (see p. 155).

Helmet: A hard space helmet won't be necessary, as the SS2 cabin acts as a giant helmet for all the astronauts.

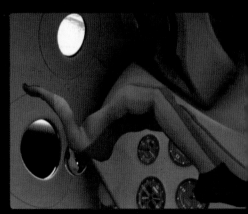

Soft soles: This conceptual artwork shows the spacesuit with soft-soled shoes for safe floating in zero gravity.

"WE'VE ALREADY GIVEN THE GO-AHEAD TO START BUILDING **MORE SPACESHIPS**... IT REALLY IS THE START OF A WHOLE **NEW ERA OF SPACE TRAVEL.**"

Sir Richard Branson, Founder, Virgin Galactic

PREPPED AND READY

At first light, *VSS Enterprise* and *VMS Eve* undergo final checks on the ground in preparation for takeoff. The vehicles are reusable. Only the rocket motor and nozzle on SpaceShipTwo have to be replaced after high altitude spaceflight. This makes space travel more afforable, and will allow many more people to travel to space.

PROFILE:
CHIEF PILOT, VIRGIN GALACTIC
DAVE MACKAY

FORMER UK ROYAL AIRFORCE TEST PILOT
Dave Mackay was born in Scotland. He joined Virgin Atlantic, and flew jumbo jets across the Atlantic for many years. The Mackay family relocated to the Mojave Desert to pursue Dave's childhood dream of going to space. He is Chief Pilot for Virgin Galactic, and is likely to pilot the first commercial flight.

DAVE MACKAY, VIRGIN GALACTIC'S CHIEF PILOT, IN HIS FLIGHTSUIT

"WHAT'S HAPPENING OUT HERE IN THE MOJAVE NOW SHOWS THAT DREAMS DO COME TRUE AND YOU SHOULD NEVER GIVE UP ON THEM."

Dave Mackay, Chief Pilot, Virgin Galactic

WET LANDING
After making a perfect landing to complete his first flight in SpaceShipTwo, Dave Mackay is sprayed with water.

Q&A

How did you become a Virgin Galactic pilot?
At the time, I was working for Virgin Atlantic as a captain on the Airbus 340. And before that, I'd been a test pilot in the [UK] Royal Air Force, after studying aeronautical engineering at university. So, I had the right background and qualifications for the job, and happened to be in the right place at the right time.

How would you sum up the experience of a flight on SpaceShipTwo?
SpaceShipTwo is a hugely exciting spacecraft, providing people with the opportunity to go into space on a thrilling rocket ride, experience the wonderful sensation of weightlessness and amazing views of Earth, the black sky of space in daytime, and the thin electric-blue atmosphere that protects our planet and provides the environment for life. For the first time, people will be able to do all this, and realize their dreams, with just a few days of simple training.

What is the highlight of the flight for you?
The rocket ride is sensational. It's where this sleek and beautiful spaceship really comes to life and shows its true capabilities—reaching over three times the speed
of sound while traveling straight up, [and] accelerating so rapidly it can reach space after a short rocket motor burn of around a minute or so.

Have you always dreamed of going into space?
Yes—has anyone not? I think almost everyone must have wondered at some time what it would be like to go into space. I know that many people, like me, have wanted to go there for a long time. Now, at last, we have a spaceship that offers the opportunity to do so to many more people than was possible before—we have a vehicle that will make the dreams of many come true.

What advice would you give to an aspiring astronaut?
Never give up on your dreams. There are now many more opportunities to get into space. Virgin Galactic will need more space pilots in the future and, perhaps, other companies will try to offer something similar, once they see how successful and popular private space travel becomes. If you want to become a professional astronaut, then work hard at school and try to stay fit and healthy. Prepare yourself as best you can so that you are ready to take advantage of any opportunity that may come along. Try to be in the right place at the right time.

How will the commercial spaceflight industry develop over the next decades?
In addition to launching people into space, Virgin will launch experiments on SpaceShipTwo. Its mothership, WhiteKnightTwo, is capable of carrying a satellite launcher, which will send small payloads into orbit around Earth. At some stage others will try to offer something similar to Virgin Galactic, and that's good—competition is healthy. In future, it may be possible to travel long distances quickly—by flying high in the upper atmosphere or near-space—going from London to Sydney, for example, in a fraction of the time it currently takes. Who knows, one day people may be able to travel to a space hotel and spend a few days orbiting Earth on a space holiday. After seeing Earth from this perspective, perhaps many will return as changed people, with an appreciation of our planet as humankind's one and only small home in the universe. A planet where political boundaries are invisible from space, a planet we have to share, and take great care of—a place that provides us with life due to the precious but

"AS SOON AS AS WE LAND, WE JUST WANT TO GO BACK AND **DO IT ALL OVER AGAIN.**"

Pete Siebold, Test Pilot, Scaled Composites

STEPPING OUT

Pilots walk across the tarmac to take their places at the controls of *VMS Eve* and *VSS Enterprise*. The cockpit of SpaceShipTwo seats a pilot and a copilot, as does the identical cockpit of WhiteKnightTwo. The pilots first learn to fly the carrier aircraft in preparation for piloting the spaceship itself. WhiteKnightTwo is a valuable training vehicle.

FLIGHT PROFILE

THE PROPOSED FLIGHT EXPERIENCE will follow set stages, starting with SpaceShipTwo (SS2) mated to WhiteKnightTwo (WK2). The vehicles climb for around 40 minutes before SS2 is released. After falling for a few seconds, SS2's rockets ignite, blasting it into space. SS2 will glide for several minutes before dropping back to Earth's atmosphere. The spaceship then heads home.

Up and down
This conceptual artwork shows the proposed flight path of the Virgin Galactic experience. After SS2 is launched mid-air, the rockets burn for around 60 seconds as the spaceship accelerates vertically. The vehicle will spiral in space for a few minutes before descending in feathered configuration for reentry into the atmosphere.

> **"...LIKE A ROLLER COASTER RIDE WHERE YOU'RE ALL EXCITED JUST TO STRAP IN..."**

Dave Mackay, Chief Pilot, Virgin Galactic, on the first stage of the flight

SS2 is released from WK2 at around 50,000ft (15.5km) and drops for a few seconds.

After gliding through space, SS2's wings de-feather and the spaceship glides back to Earth.

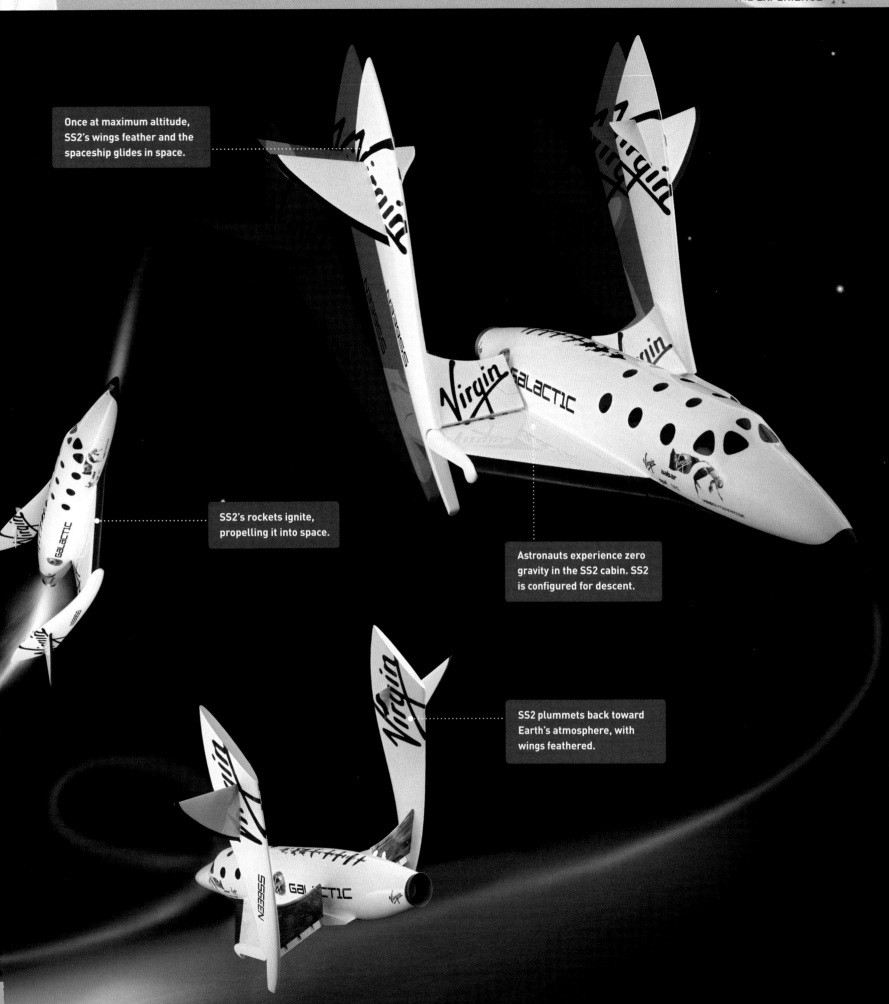

Once at maximum altitude, SS2's wings feather and the spaceship glides in space.

SS2's rockets ignite, propelling it into space.

Astronauts experience zero gravity in the SS2 cabin. SS2 is configured for descent.

SS2 plummets back toward Earth's atmosphere, with wings feathered.

THE ULTIMATE **EXPERIENCE**

WE HAVE LIFT OFF

The moment you have been dreaming of is here. Moving comfortably in your flightsuit, you climb aboard SpaceShipTwo (SS2), barely able to contain your excitement. You glance around the cabin and fasten your seat belt. The spaceship rolls smoothly on to the spaceway—the ride of your life is underway. Within seconds, you are pinned into your seat as WhiteKnightTwo (WK2) accelerates to lift off. This is it—see you soon, planet Earth!

As you leave Earth, SS2 is firmly berthed in the middle of WK2's huge wing.

① HOLD ON TIGHT
Sitting in the cabin of SS2 as it is berthed to WK2 you take a deep breath and fasten your seat belt, telling yourself that this is like any other flight.

② CHOCKS AWAY!
You hear a mechanical whirr as the chocks in front of WK2's wheels are gradually lowered and the aircraft slowly begins to roll forwards.

③ POWER UP
WK2's powerful engines fire up and as you listen to the large turbines spinning, you feel the adrenaline coursing through your body.

FLIGHT PROGRESS ···· LIFT OFF ···· SS2 RELEASED ···· SS2 ROCKETS IGNITE ···· SS2 REACHES MACH 4

THRILL RIDE

Anticipation and excitement cause the body to produce adrenaline. Your heart rate increases and you breathe more quickly. This is the thrill of a lifetime!

④ TAKING A TAXI
As WK2 eats up the 12,000ft (3,660m) spaceway, you feel a thrill of exhilaration as the spacecraft accelerates at incredible speed.

⑤ AIRBORNE
Your stomach jumps as you feel WK2 lift away from the tarmac. Gazing out of the window, you see the wheels fold away.

⑥ INTO THE DISTANCE
Looking back to Earth, you see the spaceway diminishing behind you. You sit back in your upright seat to enjoy the ride.

SS2 ENTERS SPACE · · · · · · SS2 REACHES MAXIMUM ALTITUDE · · · · · · DESCENT INITIATED · · · · · · SS2'S WINGS DE-FEATHER AND SS2 GLIDES BACK TO EARTH · · · · · · TOUCHDOWN

THE DROP

You relax as WhiteKnightTwo (WK2) climbs upwards. You take in the views of the distant desert below. Now your pulse quickens, and it's time for action again. Your seat lowers to the cabin floor and you are fully reclined. Then you feel it—SpaceShipTwo (SS2) is dropping—1, 2, 3, 4 seconds of... nothing. You are in freefall 50,000ft (15.5km) over the desert floor.

1 REACHING HEIGHTS

As WK2's gentle ascent reaches its peak, you look across for a final glance at the elegant fuselage of the mothership alongside you.

2 BE PREPARED

You compose yourself as your cabin seat slowly lowers to the cabin floor. You take a deep breath and brace yourself for the next stage of the flight.

FLIGHT PROGRESS ·········· LIFT OFF ·········· SS2 RELEASED ·········· SS2 ROCKETS IGNITE ·········· SS2 REACHES MACH 4

GOING INTO FREEFALL!

FEELING NEGATIVE?

When you fall fast, you experience zero gravity. The downward acceleration causes fluids in your body, such as blood, to move slower than solid parts, such as muscle and bone. This causes a sensation of weightlessness.

3 RELEASE
With a gentle bump, SS2 falls away from WK2. You feel your body rise against your seat belt as the g-force pulls you up.

4 FALLING FEELING
You count the seconds as SS2 drops. As you lie back in your seat, the feeling of the g-force is not as strong.

| SS2 ENTERS SPACE | SS2 REACHES MAXIMUM ALTITUDE | DESCENT INITIATED | SS2'S WINGS DE-FEATHER AND SS2 GLIDES BACK TO EARTH | TOUCHDOWN |

LEAVE
EARTH BEHIND

IT'S ROCKET SCIENCE

Rubber and nitrous oxide combine in SpaceShipTwo's (SS2) combustion chamber. The chemical reaction forms hot gases, which are expelled at high speed through the nozzle at the back, giving momentum to the engine. The thrust force you experience is the reaction of the rocket motor to the ejection of the hot gases. It's a little like a garden hose being forced backwards when the water starts flowing through the nozzle.

SS2 ENTERS SPACE .... SS2 REACHES MAXIMUM ALTITUDE .... DESCENT INITIATED .... SS2'S WINGS DE-FEATHER AND SS2 GLIDES BACK TO EARTH TOUCHDOWN

FEEL THE FORCE

The rockets of SpaceShipTwo (SS2) ignite and you are pinned back in your seat. You feel the g-force pushing from the front to the back of your chest as unimaginable power surges through the spacecraft.

1 ROCKET POWER

After the nothingness of freefall, the surge of power as the rockets ignite is overwhleming. The howl is like nothing you have ever heard.

2 UP AND AWAY

After 5–10 seconds of blast, you feel SpaceShipTwo turn upwards to climb more steeply. You look at your console in disbelief as you pass Mach 3.

FLIGHT PROGRESS ••••••••• LIFT OFF •••••••••• SS2 RELEASED •••••••••• SS2 ROCKETS IGNITE •••••••••• SS2 REACHES MACH 4

MAKING HARD WORK OF IT

Massive acceleration produces g-force, which pushes your blood back down to your feet. Your heart needs to work harder to pump blood around your body. Lying down reduces the effects, but your body feels heavy.

MARSSCIENTIFIC.COM & CLAY CENTER OBSERVATORY

③ VERTICAL CLIMB
You are now in a vertical climb towards space. You catch your breath. The g-force increases to three as you hurtle through Earth's atmosphere.

④ PINNED BACK
Excitement builds as you reach maximum acceleration in the climb. The view outside your window has turned to black.

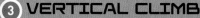
SS2 ENTERS SPACE ·········· SS2 REACHES MAXIMUM ALTITUDE ·········· DESCENT INITIATED ·········· SS2'S WINGS DE-FEATHER AND SS2 GLIDES BACK TO EARTH ·········· TOUCHDOWN

BECOME AN ASTRONAUT

You watch out your window as SpaceShipTwo (SS2) climbs higher and higher. You congratulate yourself on becoming an astronaut as the rocket hurtles you into space.

| SS2 ENTERS SPACE | SS2 REACHES MAXIMUM ALTITUDE | DESCENT INITIATED | SS2'S WINGS DE-FEATHER AND SS2 GLIDES BACK TO EARTH | TOUCHDOWN |

SPIRAL IN SPACE

The howl of the rockets cuts out in an instant, leaving an overwhelming silence. The deep blue outside the window is now an inky, infinite black. SpaceShipTwo (SS2) has reached its maximum altitude and its tails lift up into a feathered position. You are drifting in space.

① SILENCE PLEASE

The deafening roar of the rockets stops suddenly, leaving a profound silence. Your senses are reeling from the ascent as you take in your surroundings.

② FEATHERS UP

At the push of a button, SS2's wing feathers begin to lift up. You are now sitting in a shuttlecock high above the world.

FLIGHT PROGRESS

LIFT OFF   SS2 RELEASED   SS2 ROCKETS IGNITE   SS2 REACHES MACH 4

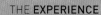

THE PILOT USES THRUSTERS TO ROTATE THE SHIP.

REACHING NEW HEIGHTS

SpaceShipTwo reaches the highest point of the flight. The engine is turned off, and the spaceship drifts through the beautiful blackness.

③ GENTLE GLIDE

Looking out of your window, you realize that the spaceship is turning in gentle spirals as it glides through space.

④ SEAT LOWERS

The seats all lower to the floor. You're going to need every inch of that space inside the cabin—you can't wait to explore it.

SS2 ENTERS SPACE	SS2 REACHES MAXIMUM ALTITUDE	DESCENT INITIATED	SS2'S WINGS DE-FEATHER AND SS2 GLIDES BACK TO EARTH	TOUCHDOWN

FEEL THE FREEDOM OF ZERO GRAVITY

The world inside the cabin has changed. There is no up or down in zero gravity. At once disorientating and awe-inspiring, zero gravity is a freedom like no other. You leave your seat and eagerly explore this alien environment, savoring every single moment.

Pushing off walls with soft-soled shoes propels you around the cabin

Soft-shell helmet protects your head as you float

❶ UPLIFTING FEELING
You feel your body begin to float and strain gently against the seat belt as the ship drifts and spirals silently through space.

❷ RELEASE
At the push of a button, you rise effortlessly out of your seat, and find that you can move more easily than you ever dreamed was possible.

FLIGHT PROGRESS········· LIFT OFF ········· SS2 RELEASED ········· SS2 ROCKETS IGNITE ········· SS2 REACHES MACH 4

WHICH WAY IS UP?

When there is no up or down, your sense of balance is thrown off. At first, you are giddy, but your brain adjusts and soon you're flying free.

③ GRACEFUL GLIDE

You push off gently and glide across the cabin. Finding yourself at a window, you gaze out in awe at the blackness of space.

④ TURN A SOMERSAULT

You tuck up your knees and spin yourself in a graceful midair somersault. The sensation reminds you of swimming.

| SS2 ENTERS SPACE | SS2 REACHES MAXIMUM ALTITUDE | DESCENT INITIATED | SS2'S WINGS DE-FEATHER AND SS2 GLIDES BACK TO EARTH | TOUCHDOWN |

LOOK BACK TO EARTH

You glide to a window and take time to drink in the view. Nothing can prepare you for what you see. The bright jewel of Earth is contrasted against the impenetrable blackness of space. You are struck by the delicate, thin blue line of Earth's atmosphere, and find yourself experiencing emotions that are hard to define.

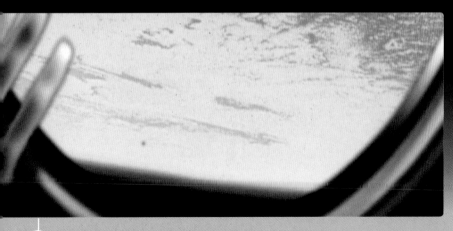

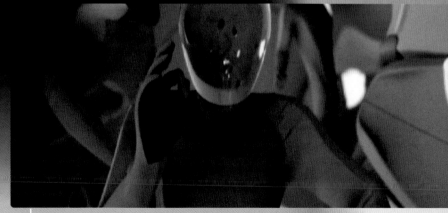

1 LOOKING BACK
Out of the window is a view that you have seen in countless images, but that you are quite unprepared for—Earth.

2 AFLOAT IN SPACE
As you float, you find your head is stuffy. Without gravity, fluids in your body are distributed evenly, meaning your blood no longer falls to your feet.

FLIGHT PROGRESS •

| LIFT OFF | SS2 RELEASED | SS2 ROCKETS IGNITE | SS2 REACHES MACH 4 |

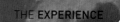

THIN BLUE LINE

The fuzzy blue line surrounding Earth that is visible from space is called the geocorona. It is the outer edge of the upper section of our atmosphere, known as the exosphere.

❸ TIME FLIES

After several minutes of exploring the weightless environment of the cabin, the signal is given for you to return to your seat.

❹ READY TO DROP

Once you are seated, SpaceShipTwo (SS2) prepares for the descent. You buckle up and enjoy in your last few moments in space.

| SS2 ENTERS SPACE | SS2 REACHES MAXIMUM ALTITUDE | DESCENT INITIATED | SS2'S WINGS DE-FEATHER AND SS2 GLIDES BACK TO EARTH | TOUCHDOWN |

FALL FROM THE STARS

Lying in your seat, you feel the ship speed downward. The g-force returns and it is stronger than before. You brace yourself as SpaceShipTwo (SS2) heads back to Earth. In a matter of seconds, you have reentered Earth's atmosphere.

① DROP DOWN
As SS2 drops like a rock from the stars, the seat straps hold you firmly in place. The view from your window changes from black to indigo.

② MAXIMUM FORCE
Picking up speed as you fall, you are experiencing strong levels of g-force. The pressure builds on your chest and you feel lightheaded.

③ REENTRY
After around 90 seconds of traveling Earthward, you pass through the atmosphere. You now have a hazy view of the desert below you.

FLIGHT PROGRESS •······· | LIFT OFF | ········ | SS2 RELEASED | ········ | SS2 ROCKETS IGNITE | ········ | SS2 REACHES MACH 4 |

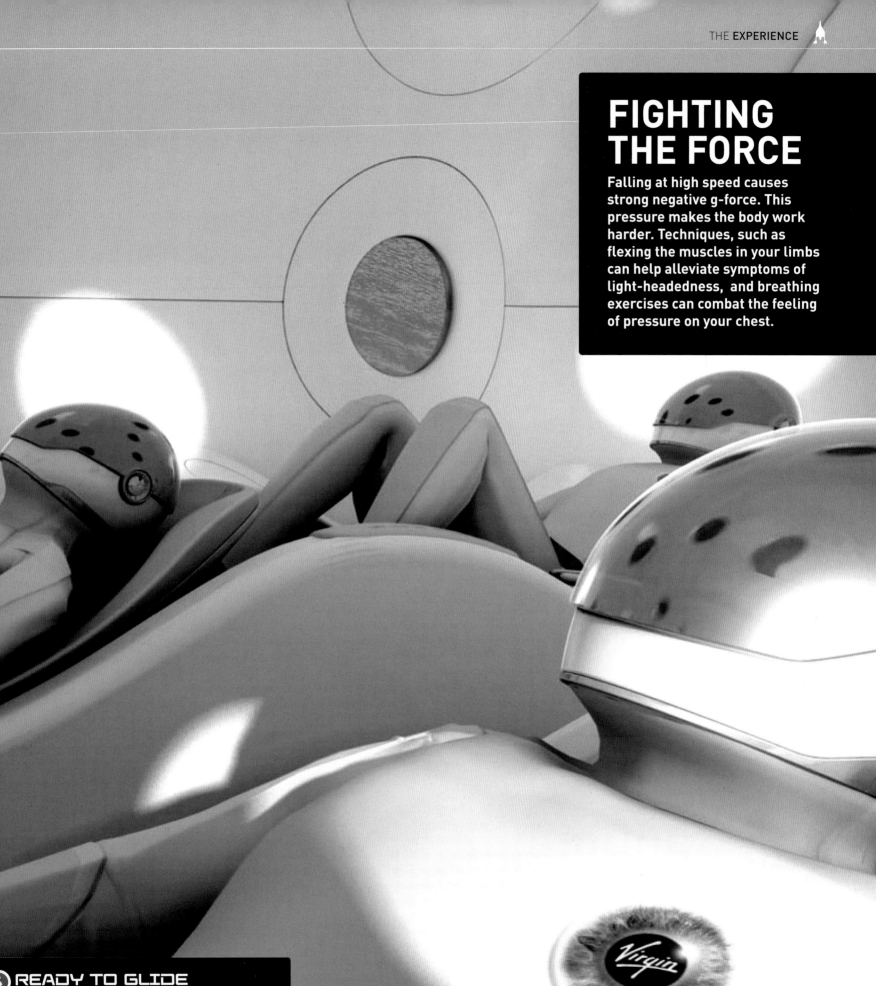

FIGHTING THE FORCE

Falling at high speed causes strong negative g-force. This pressure makes the body work harder. Techniques, such as flexing the muscles in your limbs can help alleviate symptoms of light-headedness, and breathing exercises can combat the feeling of pressure on your chest.

4 READY TO GLIDE

SS2's wings fold back from their feathered position, and the cabin is now horizontal. It won't be long until you're back at Spaceport America.

| SS2 ENTERS SPACE | SS2 REACHES MAXIMUM ALTITUDE | DESCENT INITIATED | SS2'S WINGS DE-FEATHER AND SS2 GLIDES BACK TO EARTH | TOUCHDOWN |

GLIDE BACK TO EARTH

Tensed during the thrill of your fall from the stars, you feel your muscles begin to relax. You stretch and start to reflect on your experience. SpaceShipTwo (SS2) has leveled out and your seat returns to the upright position. You look out of the window, and the sky is now a familiar blue. You're on your way home. But you know that you will never see the world in quite the same way again.

① TAKING STOCK

As your seat gently rises to an upright position, you gather your thoughts. Gazing out of the window, you see the familiar view of the desert below.

FLIGHT PROGRESS • • • • • LIFT OFF • • • • • • • SS2 RELEASED • • • • • • • SS2 ROCKETS IGNITE • • • • • • • SS2 REACHES MACH 4

UNPOWERED FLIGHT

Its rocket spent, and feathers folded back, SpaceShipTwo (SS2) becomes a glider for the final section of the flight. Once back in Earth's atmosphere, the body of the spaceship naturally tilts back into a horizontal position.

2 IN LINE

Once back in Earth's atmosphere, SS2 levels into a smooth glide. You are drifting back to Earth with no howling rocket, no engine, and no g-force.

| SS2 ENTERS SPACE | SS2 REACHES MAXIMUM ALTITUDE | DESCENT INITIATED | SS2'S WINGS DE-FEATHER AND SS2 GLIDES BACK TO EARTH | TOUCHDOWN |

TOUCH DOWN

As you look out of the cabin window, you spot the runway in the distance. Your flight is nearing its end. SpaceShipTwo (SS2) drops again and lowers its wheels. With a gentle bump, the spaceship meets the tarmac. A blur of desert flies past your window, the brakes go on, and you come to a halt. Welcome back to planet Earth!

① DROPPING DOWN
Your ears pop as the glider drops lower and lower. You hear the whirr of the wheels engaging in preparation for the landing.

② THE APPROACH
Your gentle 45-minute descent draws to an end as SS2 approaches the runway. You feel the wheels touch down on the surface.

FLIGHT PROGRESS

| LIFT OFF | SS2 RELEASED | SS2 ROCKETS IGNITE | SS2 REACHES MACH 4 |

WHEELS DOWN

SS2 approaches the 12,000ft (3,660m) spaceway at Spaceport America. Two wheels are lowered from the fuselage for touchdown, and a nose brake brings the glider to a stop.

COMING IN FOR A LANDING!

③ PUT ON THE BRAKES

As you tear along the runway, SS2 tilts forwards slightly and you feel the resistance of the nose brake as it skids against the tarmac.

④ JOURNEY'S END

As the desert speeds past your window, SS2 slows to a halt. You're glad to be home, but wish you could do it all over again.

SS2 ENTERS SPACE	SS2 REACHES MAXIMUM ALTITUDE	DESCENT INITIATED	SS2'S WINGS DE-FEATHER AND SS2 GLIDES BACK TO EARTH	TOUCHDOWN

THE DREAM GOES ON

SUBORBITAL SPACEFLIGHT IS ONLY THE FIRST STEP for Virgin Galactic. Richard Branson hopes that his company will soon be building orbital spaceships that can take his customers much further afield. A massive technological breakthrough is needed to make orbital commercial spaceflight possible. Hopefully, this dream will become a reality in the near future.

Flying further
Virgin Galactic's ultimate ambition is far loftier than space tourism. Richard Branson believes that the human race will need to utilize off-planet resources to survive. The search is on for worlds that are hospitable to humans. This could all start with a base on the Moon, where Virgin Galactic hopes to build a luxury space hotel.

DESTINATION MARS?
One day, Virgin Galactic hopes to be able to offer passengers rides to Mars and other exciting destinations in space.

THE MOON

Humans last set foot on the Moon in 1972. Getting back there would be a good next step. Before landing on the lunar surface, the big challenge is to find a cheaper, safer way of getting into orbit than the costly and high-risk methods used previously.

TITAN

Saturn's largest moon, Titan, is nearly two billion miles (3 billion km) from Earth. It is the only object in the Solar System, other than Earth, known to have stable liquid on its surface. Its dense atmosphere suggests it may one day, in the very distant future, become habitable.

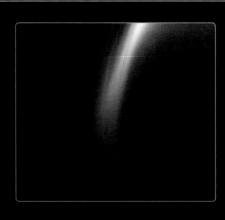

"...I'M AN OPTIMIST. WE WILL REACH OUT TO THE STARS."

Professor Stephen Hawking, theoretical physicist and future astronaut

MESSIER 33, TRIANGULUM CONSTELLATION

The flights of SpaceShipTwo are humanity's first childlike steps in commercial space travel. Hopefully, one day, space tourism will go much farther afield—from visiting the furthest reaches of our own galaxy to gazing on the distant stars of galaxies beyond the Milky Way.

INDEX

LONDON, NEW YORK, MUNICH, MELBOURNE AND DELHI

Senior Editor Tori Kosara
Editor Pamela Afram
Pre-Production Producer Marc Staples
Senior Producer Alex Bell
Art Director Lisa Lanzarini
Publishing Manager Julie Ferris
Publishing Director Simon Beecroft

Written by Ruth O'Rourke
Designed for DK by XAB Design
Jacket Design by T.W. and L.L.

Head of Brand Partnerships Neil Ross Russell
Brand & Marketing Executive Tom Westray
Vice President, Special Projects Will Pomerantz

First published in Great Britain in 2014
by DK Publishing,
345 Hudson Street, New York, New York 10014

14 15 16 17 10 9 8 7 6 5 4 3 2 1
001–262201–Sept/14
© 2014 Virgin Galactic

A CIP catalogue record for this book is available from the British Library.

ISBN: 978-1-4654-2470-9
Printed and bound by Hung Hing

Discover more at
www.dk.com
www.virgingalactic.com

"THIS TRIP
WILL BE A
TRIP LIKE
NO OTHER."

Sir Richard Branson, Founder, Virgin Galactic

The publisher would like to thank the following for their kind permission to reproduce their photographs:

(Key: a-above; b-below/bottom; c-centre; f-far; l-left; r-right; t-top)
Images courtesy of Virgin Galactic, Mark Greenberg, Tom Westray, Brian Binnie, MarsScientific & Clay Center Observatory, Jason DiVenere

12-13 Corbis: Stocktrek Images / Corey Ford (t). 12 Corbis: (fcrb); Tetra Images / Jamie Grill (fclb). Getty Images: (cb); Print Collector (clb/
Roger Bacon); De Agostini (clb/Daedalus and Icarus); SSPL via Getty Images (clb/Designs for a flying machine, crb/Balloon artwork); Time
& Life Pictures (crb/Blue Balloon). 13 Corbis: Bettmann (crb/Trans-Chanel Airplane, crb/Dr. Goddard); Hulton-Deutsch Collection (clb/Otto
Lilienthal); Stefano Bianchetti (cb); ClassicStock (fcrb). Getty Images: SSPL via Getty Images (fclb, clb/Giffard Airship); UIG via Getty
Images (crb/Wilbur Wright). 14-15 Corbis: Stocktrek Images / Corey Ford. 14 Corbis: Bettmann (clb/Amelia Earhart, fcrb); Museum of Flight
(cb, crb). Getty Images: (clb/Heinkel Plane); SSPL via Getty Images (fclb); Roger Viollet (clb/World War II. Launching site); Heritage Images
(crb/Sputnik 1). 15 Corbis: Bettmann (clb/Concorde); Reuters / Nasa TV (cb); NASA (crb/Space Station). Getty Images: (clb/BOAC Jumbo
Jet); UIG via Getty Images (clb/Soviet cosmonauts); Stocktrek (fclb). Scaled Composites LLL: (crb/SpaceShipOne). 18-19 Corbis: Bettmann.
18 Corbis: Bettmann (clb). 20-21 Getty Images: Picture Press / Detlev van Ravenswaay. 20 NASA: (cl). 21 Corbis: (br); Science Faction /
NASA (cr). NASA: (cra). 22-23 Corbis: Reuters. 23 Corbis: James Marshall (cr); Reuters (tc, tr). 24-25 NASA. 25 Corbis: National Geographic
Society / Nasa (tl/STS-26). ESA: (ftr). Getty Images: AFP (tc); SSPL via Getty Images (tr). Science Photo Library: Ria Novosti (tl/Vostok 1).
28-29 Corbis: Bettmann. 32-33 Corbis: Sygma / Jacques Langevin. 33 Getty Images: AFP (crb); Gamma-Rapho via Getty Images (cra, cr).
34-35 Corbis: Reuters / Pool / Areo News Network / Jim Campbell (t). 34 Courtesy of XPRIZE: (bl, bc, br). 35 Courtesy of XPRIZE: (bl, bc, br).
40-41 Corbis: Stocktrek Images / Corey Ford. 44 Scaled Composites LLC: (br). 45 Scaled Composites LLC: (bl, br). 46-47 Scaled Composites
LLC. 47 Scaled Composites LLC: (br). 48 Scaled Composites LLC: (bl). 48-49 Scaled Composites LLC. 55 Scaled Composites LLC. 56-57
Scaled Composites LLC: Vulcan Productions / Discovery Channel. 107 Getty Images: (cr). 129 Scaled Composites LLC: (tl). 140-141 Corbis:
amanaimages / VGL. 158-159 Corbis: Corbis Outline / Brian Smith. 192-193 Corbis: Denis Scott. 193 Corbis: NASA / JPL-Caltech / Michael
Benson / Kinetikon Pictures (tc); National Geographic Society / Kent Kobersteen (tl). 194-195 Corbis: Stocktrek Images / Robert Gendler

All other images © Dorling Kindersley
For further information see: www.dkimages.com

ACKNOWLEDGEMENTS

Dorling Kindersley would like
to thank Tom Westray